"Hate war, but love the American Warrior."

—Lieutenant General (ret) Harold "Hal" G. Moore

Vietnam War
50th Commemoration
A Time to Honor
Stories of Service, Duty, and Sacrifice

Iowa First Edition © 2017 Remember My Service Productions, a division of StoryRock, Inc. Unless credited otherwise, all photographs and copyrights herein are provided by U.S. National Archives and Records Administration (NARA).

RememberMyService.com

Vietnam50Gift.com

Library of Congress Cataloging-in-Publication Data

Vietnam War
50th Commemoration
A Time to Honor
Stories of Service, Duty, and Sacrifice
(Iowa Edition)

ISBN: 978-0-9863285-4-1

VIETNAM WAR
50TH COMMEMORATION

A TIME TO HONOR

STORIES OF SERVICE, DUTY, AND SACRIFICE

Lt. Col. (ret) E. W. "Al" and Kathleen Gardner

When the 120th Tactical Fighter Squadron of Colorado became the first Air National Guard unit sent to Vietnam, we were not surprised. Almost immediately, we left our loved ones and other pilots standing on the tarmac at Buckley Air National Guard Base in Denver as we joined our fighter group of 22 F-100s and headed west. Our thoughts were with our loved ones who were going to be without us for a year, and the group of pilots who wanted to go with us but were not selected. No one would volunteer to *not* go. We also knew the odds were against us all returning. The remainder of our officer group and 365 enlisted men were close behind us in transport aircraft.

When we crossed the West Coast, we joined with KC-135 tankers who provided our fuel and navigation. With overnight stops in Hawaii and Guam, and after 15 air refuelings, we arrived in Phan Rang Air Base, Vietnam.

As an example of our readiness, we were deployed with 100% of us operationally ready and 100% of the pilots flight-lead qualified. We were immediately integrated and flying alongside the other three F-100 regular squadrons at Phan Rang Air Base.

Our mission was to save lives. We were tasked with direct air support, close air support, and 24-hour alert to cover troops-in-contact situations. We (the pilots) flew into harm's way on nearly every mission. It was often just as dangerous on the ground at Phan Rang, where we were frequently attacked with mortars and rockets. Some of our first awards were Purple Hearts presented to injured Airmen.

It was my privilege to serve my country and to be with courageous men. No pilot would give up a mission.

The 120th became well-known throughout South Vietnam for their accuracy and professional skills, and received many commendations. We had hoped to go home without any loss of life, but we lost two of our finest in the last two weeks of our tour.

I hope the stories in this book will help you get to know these remarkable warriors. I witnessed moments of exceptional bravery in Vietnam, and I was inspired by my fellow pilots' resilience and commitment. Not all those who served have an exciting story to tell, but each has a story. Each Soldier, Sailor, Airman, Marine, and Coast Guardsman has memories of the war they'll never forget, and if you put all these memories and experiences together, you'll begin to get an idea of what it was like in Vietnam when we fought there.

My wife, Kathleen, was one of the many loved family members who suffered with the unknown while we were gone, and she would like to join me in honoring and paying tribute to those who served our country with me. We mourn for our friends who died, and for the families who lost a loved one. Their sacrifices were not in vain—they fought for freedom and democracy and helped prevent the spread of communism, and we are deeply indebted to them for that. We are proud to be a part of this legacy of service. To our brothers who returned—thank you, and welcome home.

Our friend Al Gardner, a great American, passed away January 28, 2016, before publication. We dedicate this book and documentary to Al, whose life was defined by his love of family, faith in God, and devotion to his country.

Joe and Kathleen Sorenson

Greater love hath no man than this, that a man lay down his life for his friends.
—John 15:13

President Nixon said there was no event in American history more misunderstood than the Vietnam War. Thousands of Americans hotly protested the war, and the popular media erroneously propagated many myths from the war as facts. These myths were ultimately devastating to the men and women who risked their lives fighting there. It wasn't until many years later that mainstream Americans realized the truly noble and valiant sacrifices their Soldiers had made in Vietnam.

As we recognize the 50th anniversary of the Vietnam War, we have cause to reflect on the heroism and valor of our brave American Soldiers who so valiantly served their country all those years ago. They fought and sacrificed for a noble cause. Many died for it, giving the ultimate sacrifice for freedom and for country.

With that in mind, there's hardly a group of individuals more deserving or more in need of having their story told, and told correctly. Our hope is that this book will be both a comfort and a tribute to the noble veterans of the Vietnam War. Many have spent years forging the path to recovery and resolution. We join with countless others in saluting them for their service to their God and their country.

PARTNERS

The Washington Monument and Vietnam Veterans Memorial Wall at Sunrise, Washington, D.C. *Photo by BeccaVogt via Getty Images/istockphoto.*

IOWA

The State Capitol Building situated on a hill facing downtown Des Moines. *Photo by Arias Photos via Getty Images/iStockphoto.*

To Iowa's Vietnam Veterans:

The Vietnam War was one of the most complex, misunderstood, and painful times in our nation's history. As painful as that war was, the greatest shame to come out of the conflict was the way we treated the men and women who served there. You were blamed for a war you didn't start and came home to face hostility and anger, when you should have been celebrated for stepping up to serve your country. Your caustic reception all those years ago was a national disgrace.

Despite being rejected by the very nation you sacrificed so much to serve, when you returned you continued to exemplify all that is good about America. Through your actions, you taught us all how to separate the politics of war from the courage of the American warrior. You led the movement to give new generations of veterans the welcome home you never received. Through these actions you have earned your place among America's greatest generations.

This book commemorates the 50th Anniversary of the Vietnam War by sharing stories of service, bravery, and patriotism from the many men and women who answered the call of duty. Sharing your stories will cement your legacy and help future generations of Americans understand your sacrifice and appreciate the depth of your patriotism.

On behalf of the people of Iowa, I offer you my heartfelt gratitude. Thank you for your service, and welcome home!

Sincerely,

Terry E. Branstad
Governor of Iowa

Iowa cornfields ready to harvest. *Photo courtesy of Iowa Public Television.*

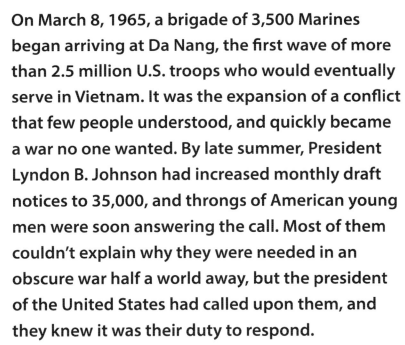

Iowa Welcomes Home Its Veterans

On March 8, 1965, a brigade of 3,500 Marines began arriving at Da Nang, the first wave of more than 2.5 million U.S. troops who would eventually serve in Vietnam. It was the expansion of a conflict that few people understood, and quickly became a war no one wanted. By late summer, President Lyndon B. Johnson had increased monthly draft notices to 35,000, and throngs of American young men were soon answering the call. Most of them couldn't explain why they were needed in an obscure war half a world away, but the president of the United States had called upon them, and they knew it was their duty to respond.

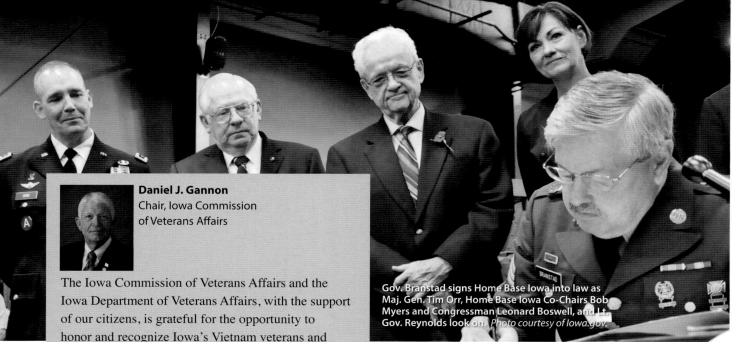

Gov. Branstad signs Home Base Iowa into law as Maj. Gen. Tim Orr, Home Base Iowa Co-Chairs Bob Myers and Congressman Leonard Boswell, and Lt. Gov. Reynolds look on. *Photo courtesy of Iowa.gov.*

The Iowa Commission of Veterans Affairs and the Iowa Department of Veterans Affairs, with the support of our citizens, is grateful for the opportunity to honor and recognize Iowa's Vietnam veterans and their families through this commemorative book. I would like to thank the following Commissioners for their generous financial support for this special project: Monica Brouse, George Goebel, Steven Hyde, Ronald Langel, Elizabeth Ledvina, Kate Myers, Gary Wattnem, and Mary Van Horn.

It is my distinct pleasure to help pay tribute to a generation of heroes whose bravery and devotion to duty have been overlooked for far too long. Thousands of Iowans willingly and proudly served in this conflict, and regardless of your role, each of you should be thanked, honored, and always remembered.

I am proud to represent an organization that stands as a voice and an advocate for veterans and their families. We are dedicated to ensuring Iowa's Vietnam veterans, their dependents, and their survivors receive the full measure of their benefits and are treated with compassion, integrity, and dignity. I would like to take this opportunity to reaffirm our commitment to you and to salute you for your service.

This book contains stories of service and sacrifice from across all different branches and locations. If your story is not in this book, I encourage you to share it with your family, friends, neighbors, and community. Our future generations need to remember your stories. Your selflessness and patriotism represent all that is good about America. God bless you all!

With deep respect and gratitude,
Daniel J. Gannon

Over one hundred thousand of those who answered the call were proud Iowans—young men and women who willingly left school, jobs, family, and friends to serve their country in a foreign land. Of that number, nearly 6,000 were wounded and 869 returned home under a flag-draped coffin, having laid, in the words of Abraham Lincoln, "so costly a sacrifice upon the altar of freedom." Thirty Iowans never returned home; their names remain on a list of those considered to be missing in action. Five of the Hawkeye State's native sons received the Medal of Honor for valorous actions "above and beyond the call of duty."

Numbers, of course, only scratch the surface. Behind the statistics are real people whom the war touched, influenced, and changed.

Gerald Berry, formerly of Des Moines, was the captain of his college football team in 1967 when he received a phone call from a father he had not seen in years. When, over lunch, Gerald asked his father the reason for his visit, he recalled his dad issuing a call to action. "Well," he remembers his father saying, "I'm a Marine from World War II. You were born on the day I crossed the beach at Iwo Jima. If there's another war, and the Marines are involved, you need to attend." Resigned to the fact that Vietnam was probably an inevitable part of his future, Gerald heeded his father's advice, joined the Marines, and arrived in Vietnam in January 1969. As he was disembarking the aircraft, he remembers a line of exuberant men waiting to return home on that same plane. "It was kind of an eerie start to [your deployment knowing] you had 13 months to go and they were just completing [their tour]," he recalls.

As a Sea Knight helicopter pilot, Berry participated in one

Wells Fargo volunteers carry a garrison-sized U.S. flag past the review stand during the Veterans Day parade at the Iowa State Fair.
Photo by Rodney White / The Register.

Thank you for your support!

A special debt of gratitude to Dave Miller, Senior Producer for Iowa Public Television, whose assistance and resources were indispensable in making this publication possible. We also wish to thank the Iowa Veterans License Plate Fund, Iowa County Veterans Association, Iowa Public Television (IPTV), Theisen Supply, Casey's General Stores, Prairie Meadows Casino Racetrack & Hotel, Disabled American Veterans (DAV) Department of Iowa, DAV Des Moines Chapter 20, VVA Iowa State Council, Brownwells, Fareway Stores, Wells Fargo, and HyVee, Inc.

mission requiring him to insert a patrol of reconnaissance Marines deep in enemy territory. Expecting to retrieve the patrol in a few days, the helicopter crew was surprised to hear via radio that the patrol had been spotted by NVA soldiers and needed immediate extraction. By the time the Sea Knight arrived at the drop position a mere three minutes later, the men on the ground were already taking heavy enemy fire. The Cobra gunships flanking the Sea Knight depleted their ammunition, providing covering fire while Gerald maneuvered through the NVA bullets and pulled the recon Marines aboard. He was the only person on the helicopter patrol or recon team who escaped serious injury or death. "You always wonder why you were the one who didn't get touched," he remarks.

Larry Spencer, of Earlham, was serving as a radar interceptor in the U.S. Navy when he was shot down over North Vietnam in 1966. His mother received word that he was reported captured, but had to live with uncertainty about the fate of her 25-year-old son for nearly two years before that report was verified. As a prisoner of war, Larry remembers, "We represented a bargaining chip for the Vietnamese." He recalls how, at the end of each year, the prisoners

optimistically told themselves that they would be home for the *next* Christmas. The optimism waned after the first couple missed Christmases, and as this cycle of intermittent hope and despair dragged on for Larry and his fellow POWs. The mental discouragement was, at times, overwhelming. But the prisoners gained emotional resiliency within the walls of the "Hanoi Hilton."

"You really do learn that the things over which you have no control, there's not much gained by worrying about them," Larry recalled years later. "And we were really in control of very little. Sometimes the only control you had was how you dealt with it." Larry spent seven years of his life as a prisoner of war, where he learned from brutal experience the importance of hope and persistence, and the power of the human spirit. Above all, he says his multi-year tenure in a POW camp taught him "to keep the faith and keep trying. The only way you know for sure you're not going to accomplish something is to give up."

Retired Air Force Colonel Harold E. Johnson spent six years in the same infamous prison compound. The Blakesburg native and electronic warfare specialist had volunteered for a series of attack-and-destroy missions against Vietnamese

Jodi Tymeson
Commandant
Iowa Veterans Home

Since 1887, the Iowa Veterans Home has proudly served Veterans and their spouses by providing superior care and an exceptional quality of life. The Iowa Veterans Home (IVH) was established to care for our Civil War Veterans, and the cornerstone of our flagpole reads, "Iowa Forgets Not the Defenders of the Union." Today, IVH serves a larger number of Vietnam Veterans than any other war period, and we still remember and honor the defenders of our great nation. This book is a reminder of a war that was fought 50 years ago by young men and women who were dedicated to the ideals of freedom and liberty. We owe them a debt of gratitude. We owe them our heartfelt thanks. We owe them the warm Welcome Home that many of them did not receive. Vietnam Veterans, thank you for leaving your homes and families to accomplish what was asked of you. From all of us at the Iowa Veterans Home, thank you for your honorable and faithful service, and Welcome Home!

Sherrie Colbert
Museum Director
Iowa Gold Star Military Museum

I was four years old when my oldest brother, Michael, was drafted and fought in the Vietnam War. I will never forget the sadness in my mother's eyes when my brother left, and the relief when he returned home.

Our family was blessed that my brother returned home with minimal injuries, mental or physical. But there were many who served that suffered long-term effects and illness and still struggle today, and then there are those we can never forget—those who paid the ultimate sacrifice.

Thanks to the Gabus Family Foundation and Gene Gabus, the Iowa Gold Star Military Museum has a new exhibit honoring our Iowa Vietnam Veterans, telling their stories and honoring their sacrifices. Our Iowa veterans would tell you they served because it was the right thing to do, but at the Iowa Gold Star Military Museum, we are honoring Vietnam Veterans with the accolades and honor they rightly deserve.

missile sites aboard an F-105 "Wild Weasel." The crews who flew in these top-secret specialized jets had only a 50% survival rate. By April 1967, Johnson had beaten the odds by flying 92 missions over North Vietnam. On his 93rd mission, just seven shy of the 100 flights that would allow him to go home, his aircraft was shot down. Johnson and his pilot were captured a short time after parachuting to safety.

Johnson was tortured and interrogated virtually every day for one year until he exhausted his value as an intelligence source—at that point, the interrogations ceased, but the torture continued. Finally repatriated to the United States in March 1973, he continued to serve in the Air Force for 14 more years before retiring back to Blakesburg to spend time with his wife, three children, and eight grandchildren.

The stories and sacrifices of these great Americans are representative of so many brave and devoted Iowans who fought in Vietnam. And while many Iowans served in-country, the Hawkeye State's involvement in the Vietnam War extended far beyond the battlefield. On the home front, Iowa farmers developed a special variety of corn used widely in foreign aid programs across Southeast Asia. The Iowa Ordnance Plant in Burlington geared up for war production in 1966 and assembled missiles used to support war efforts. And Collins Radio of Cedar Rapids created new technology used to improve both military and civilian aircraft.

When the time came to make the journey home, many of those who served overseas returned with little to no fanfare or thanks. Certain Veterans can recall happy experiences that renewed their faith in the country they had sacrificed so much to serve. Gerald Berry tells of how, when his car experienced mechanical problems in the States, generous mechanics at a local dealership fixed and detailed his car and then refused money to show their thanks for his service in Vietnam. "If those [experiences] are what you take away," he observes, "America's a pretty good place." But the war had brought much social turmoil to a country now divided and bitter, and that bitterness caused many to treat Veterans as though the war and its atrocities were their fault.

After arriving in California, Roger Beau of Dubuque

remembers traveling to a commercial airport eager to board a plane bound for Iowa. As Roger exited his cab, he faced a mass of protesters screaming epithets who began spitting on him and hitting him with protest signs. He remained shaken by the event until his next stop, where someone from the USO recognized him as a GI and led him to a quiet room with a cot where he could get some much-needed rest. As he flew toward home the next day, he realized he was over Iowa and was overcome with emotion. "I felt like I was flying into the Garden of Eden."

Robert Myers is a decorated Vietnam Veteran, retired Army Colonel, former president and CEO and current Chairman of the Board of Casey's General Stores. He notes that while Vietnam Veterans like himself and Roger Beau did their duty as well as any other generation of American warrior, "when we returned home we were shunned, spat upon, threatened, yelled at.... It's an understatement to say we were not treated well." He and his fellow soldiers were discouraged by the Army to talk about their experiences in Vietnam, and he remembered the conflict being conspicuously absent in his military school curriculum. "It wasn't until years later," he notes, "that attitudes changed."

As time passed and the country turned its attention to Watergate, the Iran hostage crisis, inflation, and other pressing events, Americans in many ways acted like the war had never happened. Instead, they swept Vietnam under the carpet, burying it in the attic of the country's collective memory. It wasn't just that Vietnam Veterans were unappreciated—their service and experiences

Jim Theisen
War Veteran
and Civic Leader

Born in Dubuque in 1934, Jim enlisted in the U.S. Army in 1954 and spent 19 months in France and Germany as a supply sergeant. After his honorable discharge from the Army, he returned home to help his parents build Theisen Home & Auto, which now has 21 stores located across Iowa. Through the years, Jim has been quick to support the military, and recently helped finance the 2007 building of the Veterans Plaza in Dubuque.

"It is with great pride that I offer my support to this commemorative project. Our Vietnam Veterans and their families have sacrificed greatly in defense of freedom and liberty, and I'm honored to play a small part in recognizing them. Vietnam Veterans have been forgotten for so long; this gift will go a long way in ensuring that the legacy of their service lives on forever."

> "We know the pain of the ultimate sacrifice others have paid for us. We have seen it. We have lived it."
> —*Robert Myers*

went unacknowledged. They were invisible. And as Iowans moved on through the subsequent decades, so too did the Veterans themselves. They hid their emotional wounds and tried to get on with the difficult business of everyday life.

George Everett "Bud" Day was one of the Veterans making that transition. The Sioux City native had already served in both World War II and Korea when he volunteered for a tour in Vietnam in 1967. Assigned to the 31st Tactical Fighter Wing at Tuy Hoa Air Base, Major Day's extensive flight hours and piloting experience got him a job commanding a detachment of F-100 Super Sabres that flew missions in high-threat areas. In August 1967, his aircraft was shot down and he was taken prisoner by the North Vietnamese. Finally released after five years and seven months in captivity, he was awarded the Medal of Honor by President Gerald Ford in March 1976, flanked by his wife and four children. Senator John McCain, who shared a cell with Day while the men were in captivity, remembered him at his 2013 funeral as "the bravest man I ever knew."

Former Iowa Congressman Leonard Boswell, who served as a helicopter assault pilot in Vietnam, believes the mixture of pain and patriotic pride felt by Veterans lingers long after the war has ended. To this day, "when the flag goes by, it makes my heart beat a little faster," he says. "But there are a lot of memories about the people I served with who didn't come back."

Iowa Vietnam Veterans were granted a small semblance of healing on Memorial Day 1984, when the Iowa Vietnam

Iowa Vietnam War Memorial Wall on State Capitol grounds. *Photo courtesy Iowa Public Television.*

Veteran War Monument was dedicated. Located on the lawn of the state capitol, the monument echoes the design of the national Memorial Wall, and is dedicated to all Iowans who served during the war. Entitled "A Reflection of Hope," the stone monument is engraved with the names of the 869 Iowans who made the ultimate sacrifice in Vietnam.

By the start of the Gulf War, over half a decade after the dedication of the state monument, a national awakening seemed to occur, bringing an increase of gratitude for the warriors who answered their nation's call to duty. The public support and warm welcome given to Gulf War Veterans was a sharp contrast to the cold greeting Vietnam Veterans had received just two decades prior. But none of the older Veterans felt any animosity toward the younger generation, instead resolving to never again allow one generation of Veterans to abandon another. "The bond that formed among us in combat many years ago transcends generations," comments Robert Myers. "It's not just our generation. It's all generations of Veterans. We know the pain of the ultimate sacrifice others have paid for us. We have seen it. We have lived it."

Thus, the movement to welcome home Vietnam Veterans has come, started at least in part by the Vietnam Veterans themselves. As Americans watched Vietnam Veterans welcome home their comrades-in-arms—from the time of the Gulf War up to the recent Global War on Terror—they learned by example how to separate the politics of war from the courage of the American warrior. In the opinion of Mark Smith, a native of Buffalo Center who served both in Vietnam and again in Iraq in 2004, Americans have taken that lesson to heart. "When you came back from Vietnam, you didn't tell anyone you were there, other than family and close friends. When I came back from Iraq, there were people at airports applauding and welcoming [us] home."

To date, 42 states have designated a "Vietnam Veterans Day" to commemorate the sacrifices made by Vietnam Veterans and their families and to honor the men and women who were denied a proper welcome when they returned home so many years ago. Most states celebrate the day on or around March 29—the day in 1973 that the last combat troops were withdrawn from Vietnam and the last prisoners of war held in North Vietnam arrived on American soil. In 2005, the Iowa Legislature officially designated May 7 as "Vietnam Veterans Recognition Day" statewide with the adoption of SR 139. Designation of such a day offers an opportunity for Iowans to give the estimated 76,000 Vietnam Veterans who live in the state a reception befitting the magnitude of their service and sacrifice. After all, observes Larry Spencer, "remembering is very important in helping us make a better tomorrow."

The courage and patriotism of our Vietnam Veterans—who have asked for so little despite giving so much—merits a solemn obligation of gratitude and remembrance and the embrace of a warm welcome home from the American people. Iowa has begun that welcome, and its officials and citizens have committed to honor the legacy of these warriors by word, deed, and through everlasting appreciation. ∎

Soldier communicates with approaching helicopter delivering supplies to a fire base. *Photo courtesy of NARA.*

Private First Class Joseph Big Medicine Jr., Company G, 2nd Battalion, 1st Marines, a Cheyenne Native American, writes a letter to his family back home in the United States during a mission east of An Hoa. *Photo courtesy of NARA.*

TABLE OF CONTENTS

U.S. Soldiers draw sniper fire while on a search mission.
Photo courtesy of NARA.

To Honor and Remember

Millions of words have been written about the Vietnam War: histories, memoirs, and novels that have summarized, analyzed, and dramatized, trying to untangle a difficult time in our country's past. This book sets out simply to honor the men and women who served.

They were grunts and pilots, clerks and cooks, sailors, soldiers, technicians, mechanics, privates, generals, the gung-ho and the reluctant, the bold and the ones who were afraid but served anyway. More than 58,000 of them lost their lives in the heat, wet, and mortar fire of Vietnam. The rest came home, often to cold shoulders and heated arguments from a public that had turned increasingly against the war.

Newly returning vets were often advised to change into civilian clothes as soon as they landed. Let your hair grow out, they were told. Don't talk about the war. So a generation of men and women came home and tried to fit back into a country and a time that sometimes seemed alien. Some put their medals in a closet. Many kept quiet about what they had seen and done thousands of miles away—about the pain and the camaraderie, the insanity and the valor. They came back home—some of them flourished, and some of them have never recovered.

We want to honor them all.

It has now been 50 years since the war began. Google "Ho Chi Minh" and you'll find advertisements for discount flights to a country that has changed. A half-century is too long to ignore the men and women who put their lives on hold to fulfill their duty.

The stories on these pages are heartbreaking and healing, sometimes frightening and sometimes funny, told by men and women from every state and every branch of service.

Because the Vietnam War—from the days of American advisors in the mid-1950s to the last helicopter of evacuees that took off from Saigon rooftops 20 years later—spanned such a long period of time, and was fought in such disparate conditions, the war one person remembers is often not the same as someone else's. But we hope veterans will find their history in these pages.

And if it was your spouse or parent or grandparent who served in the Vietnam War—if you have found this book on a coffee table and have opened it up, curious about what's inside—we hope you will begin to understand what the war was like and what your loved one offered this country.

Sailors load ordnance onto a fixed-wing aircraft. *Official U.S. Navy photo courtesy of Russell A. Elder.*

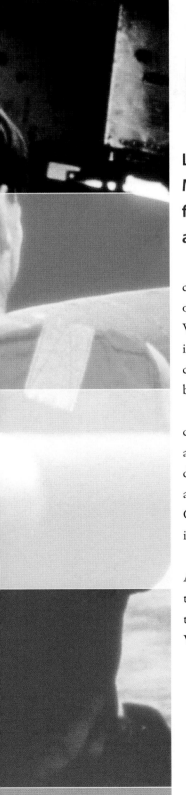

1964–1966
Early War

Long before there were 58,000 names there were just two: Major Dale Ruis and Master Sergeant Chester Ovnand, the first Americans to die in what would later become known as the Vietnam War.

It was 1959, five years after the Geneva Accords had divided Vietnam into two countries—the Republic of Vietnam (RVN) in the south, and the Democratic Republic of Vietnam (DRV) in the north. By then, the U.S. had been sending advisors to South Vietnam for four years, helping to create a South Vietnam army to fight Viet Cong insurgents led by Ho Chi Minh, who sought to unify all of Vietnam under communist control. For Americans back home, though, Vietnam was just a headline buried in the back pages of the newspaper.

But all that was about to change. Against a backdrop of fear about the spread of communism worldwide, and especially in Southeast Asia, buzzwords like *domino theory* and *escalation* crept into American conversations and newscasts. President John F. Kennedy deployed helicopters, bombers, and U.S. Army Special Forces advisors to Vietnam, and approved the use of Agent Orange and other defoliants. By the end of 1964, after the Gulf of Tonkin Resolution gave President Lyndon Johnson authorization to step up involvement, U.S. troops numbered 16,000.

In 1965, the first U.S. Marine and Army ground combat units were deployed, and the Air Force began Operation Rolling Thunder's sustained bombing of North Vietnam. By the end of 1966, almost three million Army, Air Force, Navy, Coast Guard, and Marine troops were committed to the fight, whether Stateside, abroad, or in theater. The war in Vietnam was now front-page news.

The nuclear-powered aircraft carrier USS *Enterprise* (CVAN-65) cruises in the Gulf of Tonkin off the shores of Vietnam. With A-4 Skyhawks on her bow, *Enterprise* recovers one more on her angled deck. *Photo courtesy of NARA.*

TIMELINE 1964–66

February 2, 1964
G.I. Joe makes his debut as a figure" toy.

August 7, 1964
U.S. Congress passes the Gu
Tonkin Resolution, giving Pr
Lyndon B. Johnson the autho
action against North Vietnam

December 14, 1964
U.S. begins bombing North

March 2, 1965
Vietnam War escalates with (
Rolling Thunder bombing ra
North Vietnam.

March 8, 1965
First American combat force
U.S. Marines land in Vietnan

March 21, 1965
Martin Luther King Jr. and tho
civil rights demonstrators begi
mile march from Selma to Mo

August 11, 1965
Race riots erupt in Watts sect
Angeles, California.

November 27, 1965
Tens of thousands of Americ
demonstrate against the Vietr
Washington, D.C.

December 1965
180,000 U.S. troops in Vietn

August 6, 1966
President Johnson increases
troops stationed in Vietnam t

September 8, 1966
TV program *Star Trek* premier

December 31, 1966
385,300 U.S. troops in Vietn
6,143 killed in action.

There at the Very Beginning

August 2, 1964, started out as a routine night shift at the Naval Communications Station on the Pacific island of Guam. Eugene Mathis, manning the radios, had no idea he was about to witness the beginning of the Vietnam War. "We were pretty laid back, just sitting around drinking coffee, and listening for messages from ships," recalls Mathis, then a 19-year-old Radioman 3rd Class. "The messages were in Morse code—you had your earphones on, typed out the messages on a sheet of paper, and routed them to the proper designee."

Then came the message that changed everything. "Suddenly we hear some code we haven't heard before. Someone is calling us up with an emergency message, a 'Z precedence' message," Mathis says. "I jumped up, ran over to my receiver, called the guy back up and asked him to send the message. It was a guy on USS *Maddox*, a destroyer, sending a message that they were being fired on by North Vietnamese PT boats."

The message, which was not encrypted, took Mathis by surprise. "I thought, what in the world is going on? We didn't think about anything going on in Vietnam. We knew there were advisors there, but this shook us up." He finished taking the message, "and we routed it to the proper people. The rush came at daylight, when all the officers started coming in. It was chaotic in the radio room. People were constantly coming and going—everyone was talking about it."

Mathis could not have predicted the effect that message would have on America's history. The attack became known as "the Gulf of Tonkin incident," and on August 10, 1964, Congress enacted the joint Gulf of Tonkin Resolution, giving President Johnson authorization to use conventional military force in Southeast Asia. Even without a formal declaration of war, the Johnson Administration now had the authority to begin rapidly escalating U.S. military involvement in North Vietnam.

For Mathis, this meant that relaxed shifts in the radio room were a thing of the past. "Because of its location, Guam was the hub, and a lot of [message] traffic came in to us," says Mathis. "Every time you got anything greater than an operational message, you had someone looking over your shoulder. Before this, we could leave and go to the chow hall for meals. But we were working so many hours that they would get the food and bring it in to us. It was a lot of pressure for a 19-year-old kid."

He was glad to get away from the stress when his assignment on Guam ended that November. His next assignment was the radio room on board USS *Talladega* (APA-208), which carried troops to Vietnam. "Our ship landed the first battalion of Seabees at Da Nang in 1965 to build the airstrip," Mathis says. "So I was there at the very beginning of the war, and I saw the buildup of the war firsthand."

— Eugene Franklin Mathis Jr.
U.S. Navy, Radioman 3rd Class, Naval Communications Station,
USS *Talladega* (APA-208)
Guam: 1964; Vietnam: 1965–66

Aircraft launch from USS Coral Sea (CV-43) in the Gulf of Tonkin. *Official U.S. Navy photo courtesy of Russell A. Elder.*

A newly modified Fairchild C-123K Provider on a resupply mission over South Vietnam.
Photo courtesy of NARA.

Landing Hard

The mission was simple—and dangerous. Ten Soldiers were holed up, under enemy fire, on a small Special Forces base northwest of Saigon. Air Force pilot Carl Bradford Johnson and his C-123 crew were assigned to get them out.

"The best time to do it was at night," he recalls, "so we had to do it with a blackout landing, no lights on the field. We flew what we thought was the downwind, and the airplane would shake every once in a while from the war going on under us."

The men below had set up two jeeps, each at one end of the runway, and on a signal, they would flash the headlights very quickly so Johnson could get his bearings. "It was a dirt runway, and it was wet," remembers Johnson. "It was terrible weather. It was not comfortable."

That was an understatement. Johnson gave the signal and saw the lightning-fast flash of the headlights below. "We knew that as short as the runway was, and being wet, we were going to have to put these propellers in reverse in the air," he says. "And we were going to land standing on the brakes."

And that's pretty much how it happened. They came down low and fast, touched down, put the engines in reverse, and stood on the brakes. When the plane touched down, it hit a hole in the runway, knocking the [wheel] gear doors completely off the plane. "We skidded around, and finally, we saw a flashlight waving to us. We taxied down, turned in, and we lined back up again [on the runway] because we knew we just might get a second's warning to get out of there."

The plane's engines remained on as the Soldiers on the ground loaded valuable supplies and a 105mm howitzer onto the plane as quickly as they could. Johnson helped, then heard a voice demanding to know "who the blankety-blank" was flying the plane. Johnson looked around and saw an Army sergeant, covered with mud. "All you could see were his white eyeballs and white teeth," Johnson recalls.

He grabbed Johnson and hugged him, saying, "I love you. I never thought I'd say that to an Air Force officer in my life, but I love you." The sergeant had been in one of the jeeps, and when Johnson landed, he thought the plane was going to hit him. He dove out of the vehicle, into the mud, then he lay in the mud and laughed because he was alive—and help had arrived.

Johnson and his co-pilot finished loading up the plane. "Then we lined up and released the brake," Johnson says. "And there wasn't a soul on that base when we left. It was empty."

— **Carl Bradford Johnson**
   U.S. Air Force, Captain, pilot, 315th Air Commando Wing, Troop Carrier, 2nd Air Division (Combat Cargo)
   Vietnam: 1966–67

A typical busy Saigon street scene in the 1960s. *Official U.S. Navy photo courtesy of Russell A. Elder.*

Members of the 25th Infantry Division advance along a stream.
Photo courtesy of NARA.

Fishing Line Booby Trap

Peter Nevers's squad was guarding a bridge north of Da Nang, patrolling in tall, thick, almost razor-sharp elephant grass, when the point man leading the squad suddenly froze.

"What's the problem?" Nevers asked. Not daring to move, the Marine tensely responded, "I've got something caught on my boot." Nevers moved the rest of his men back. "What does it feel like?" he asked. The Marine said it was a piece of clear fishing line. Nevers got on his hands and knees and crawled forward to look at the Marine's boot. "And sure as hell, there's a piece of fishing line caught on one of the eyelets of his boot," he recalls.

Nevers followed the fishing line where it led, and saw that the Viet Cong had cut a clump of elephant grass in half at the base, peeled it open, put in "the biggest grenade I ever saw in my life" inside, then tied it up. The fishing line was placed to trip the grenade. If the Soldier hadn't stopped when he felt the tug of the line on his boot, the whole squad would have been blown up.

But trying to detach the fishing line from his boot without tripping it would be a delicate operation. "So I tell the kid, 'Don't move—I'm going to try to get it off your boot.' He's freezing right where he is anyway." Nevers very gently removed the line off the boot, and told the Marine to move away.

"Meanwhile, the company commander is back there and wants to know why we're not moving," Nevers recalls. "I said, 'Because we've got a booby trap here, you know.' He said, 'Well, get it out of there!'"

Nevers took a parachute cord, tied it to the fishing line, and yanked—but the cord snapped. He tried again with the parachute cord, and again it snapped. "So I took a grenade, pulled the pin, put it down next to the booby trap, let it [the handle] go and ran like hell," Nevers says. "You get four seconds' delay time. At the four seconds, the thing went off. Thank God it went off straight up, so you've got an area where the shrapnel isn't coming."

With such hidden dangers all around, Nevers said prayer was the best good-luck charm.

"You get kind of religious when you get into those situations. You rely on each other and pray a lot."

— **Peter J. Nevers**
   U.S. Marine Corps, Sergeant, squad leader, Company G, 2nd Battalion, 26th Marines, 5th Marine Division
   Vietnam: 1966–67

Gotta Do What We Gotta Do

Pilot Walter Paulsen's main responsibilities in South Vietnam were two-fold—he transported ARVN troops to assigned locations and he worked on the aircraft maintenance crew. His expertise as a pilot was invaluable in maintaining and repairing the fleet of 25 UH-1B Huey helicopters in his company.

But despite all the work on his plate, Paulsen didn't hesitate when a major tapped him on the shoulder and said, "Lieutenant, I just received a radio call from some [U.S.] advisors at an outpost under siege, and they have about two hours' worth of ammo remaining. Can we take more in to them?" Paulsen immediately agreed.

He and his crew loaded up a helicopter with ammunition from a nearby outpost and took off, although they had no back-up or armed escort. "I came over the drop area at about 3,000 feet, chopped the throttle, dumped the nose, and went into a 3,000-foot-per-minute rate of descent," he says. "I pulled in the power, came to a shuddering hover, kicked out the ammo, and then got the hell out of there. I knew that someday [the enemy] might get us doing things like this, but we just gotta do what we gotta do."

In fact, it was Paulsen's efficiency—along with a reward of a jeep trailer filled with ice and beer—that eventually got him shot. "Typically, any company would have about two-thirds of their choppers flyable at one time," he explains. "One day our major said, 'The day you get all 25 legally flyable, I'll fill that trailer with ice and beer.' Well, it didn't take but a few weeks, and we got there." The major was true to his word, and filled up the trailer with ice and beer.

The crew kept all 25 helicopters flyable for a full month, and received a Meritorious Unit award as a result. Since the company had its full slate of Hueys available, it ended up "loaning" them to a nearby aviation company. But on the day Paulsen flew a mission for the other company, his Huey got hit.

"When I heard the first burst I knew it was .50-caliber, and I put the chopper in a hard right climbing turn," he recalls. "The next burst caught us with one round coming up through the floor, under and through my leg, hitting the collective control and then hitting [friend and copilot] Jack in the face. The ship yawed as it went into autorotation, but we hadn't lost engine or transmission oil pressure. We put out our Mayday and then realized there would be no help at our location before dark—we had to fly to a safe landing area."

The two injured pilots flew at about 1,000 feet in search of safety. They found an unfamiliar, rough landing strip outside enemy territory and did their best to land the damaged Huey. The two were flown to the Saigon Air Force Dispensary, where X-rays revealed that Paulsen's leg injury was actually more serious than Jack's wounds. "If I hadn't been part of the team that substantially increased our number of flyable choppers," Paulsen points out, "I wouldn't have been there that day."

— **Walter Paulsen**
  U.S. Army, 1st Lieutenant, 114th Air Mobile Company
  Vietnam: 1964

A-1E Skyraider aircraft of the 34th Tactical Group, based at Bien Hoa, fly in formation over South Vietnam on way to target. *Photo courtesy of NARA.*

Dog and handler, both treated by medical personnel. *Photo courtesy of W. Greg Nelson.*

Respecting the Native Culture

By day, the mission of the 1st Battalion, 7th Marines was to stabilize the South Vietnamese village of Binh Nghia, a peninsula community of fishermen and farmers who struggled to survive and get along in a land of sand dunes and mangrove swamps. By night, their mission was to fight the Viet Cong, some of whom were the same villagers the U.S. Marines were helping during the day.

"The Viet Cong and the South Vietnamese were from the same population," says Vince McGowan, a staff sergeant in the 1st Battalion. "There is no question they were. It was civil war. We'd say, 'Who's going to ambush who first? Who's going to walk into whose trap?'"

McGowan remembers one night he captured a Viet Cong fighter with a familiar face. "Having shot my barber, it was hard to get a good shave after that," he says. "The guy walked into our ambush. It didn't kill him, and we had him medevaced out."

McGowan says he has mixed feelings about the Vietnam War. Politicians weren't sure what to do, and the public blamed returning servicemen for the unpopular war. Even now, most American textbooks devote little space to the war, McGowan observes. But he has no reservations about the importance of his regiment's "day job," helping the people of Binh Nghia improve their lives and community.

"The idea was to give them their own form of democracy," McGowan explains. "They had to figure it out. [The government held] elections there, and people organized into different groups." The Marines advised villagers and helped elected South Vietnamese officials set up a functional government that could handle local issues like disagreements about property and personal matters. The key to success, McGowan observes, was not to direct, but only *suggest* while respecting the native culture—but it was often frustrating because "the average [Vietnamese] doesn't get that. You can't impose your beliefs on someone who doesn't know what you're talking about."

McGowan has returned to Binh Nghia since the war. "There's a marker there, thanking us for what we did," he says. All the villagers he knew as a Marine were killed after the United States withdrew its troops, but McGowan found a way to honor them and help current residents. He donated $10,000 to the village, asking that it be used to provide an area where the families of the late Popular Force* members could live in perpetuity. "That $10,000 isn't much," McGowan says. "It's not enough to build. It's not a lot when you spread it around. But it was something I wanted to do."

— Vince McGowan
U.S. Marine Corps, Staff Sergeant, 1st Battalion, 7th Marines, 1st Marine Division
Vietnam: 1966–68

Local South Vietnamese loyal to the government who joined part-time militia to protect their own village or district. They were often referred to as "Ruff-Puffs."

The Dogs of War

In 1962—when the war in Vietnam was barely making headlines in the States and the number of U.S. troops on the ground was still in the low thousands—Greg Nelson was deployed on an urgent top-secret mission: training war dogs.

With three days' notice, Nelson and four other U.S. veterinarians deployed to Vietnam to train the kind of smart, alert dogs that could serve as scouts and sentries against enemy forces. Pretty soon, he says, the Viet Cong came to fear the war dog program. The VC relied on "the ambush, the hit-run-disappear tactic," Nelson says, but the scout dogs were able to root the enemy out of caves, holes, and from under riverbanks.

The Vietnamese veterinary technicians Nelson worked with were good students, he says, and the dogs they trained did very well. Although Nelson and his group tried using various kinds of dogs, some were unsuitable—Dobermans became indiscriminately aggressive, and Boxers had trouble in water. The best choice was German Shepherds, imported from Germany.

At first, "the dogs were scared to death of the Vietnamese and would break away from them and run to us for protection," Nelson remembers. "They had never seen Asians and didn't understand their language. Later, the dogs gave their complete allegiance to the Vietnamese handlers and were wary of Caucasians."

— W. Greg Nelson
U.S. Army, 1st Lieutenant, war dog trainer, Veterinary Corps
Vietnam: 1962

The Code

In 1965, North Vietnamese forces shot down Smitty Harris's fighter jet near Than Hoa, then took him to Hanoi and threw him into a dark, dank cell. He was the sixth American prisoner of war to be held captive in what the POWs called the "Hanoi Hilton."

After a few months, Harris was put in a larger cell with three other POWs. That first night, the four of them whispered until dawn. Not long after that, Harris taught them the "tap code." He had learned it almost by accident when he was in survival training at Stead Air Force Base. "Here, I'll draw it for you on the blackboard," the instructor had said as an afterthought. The code was simple: five letters across, five down, a first tap signifying the horizontal row, the second tap the vertical.*

The tap code turned out to be a lifeline for hundreds of American POWs. "It was the most important thing that kept us together," says Harris. Through the code, tapped out on the walls of their cells, the POWs passed along information about who was being tortured, who had arrived, and any scrap of world news from the arrivals. They maintained military chain of command and tried to keep each other's spirits up. "It gave us a sense of unity and pride," he says.

When their captors eventually realized that the tapping noises were a means of communication, they stationed guards between the cells. So the prisoners coughed out the letters of the code. They got down on their knees and flashed the code with their hands, under the space between their doors and the floor. They took their purple diarrhea pills, crushed them up, spat on them and made ink to write messages on toilet paper.

And when the North Vietnamese tortured them long enough—making them kneel for days, tying them up in a ball, beating them—the prisoners would agree to write statements extoling their captors, which the North Vietnamese tried to use as propaganda. But even then, the prisoners found a way to communicate what they really meant. One prisoner agreed to write that pilots in his squadron were so anti-war they refused to fly—but he said the pilots were Dick Tracy and Clark Kent.

Harris was a POW for 2,871 days. When he returned home and saw his daughters for the first time, they ran toward him, squealing, and jumped in his arms. *Oh thank you, Lord, they haven't forgotten*, Harris remembers thinking. He also met the son who was born not long after Harris's plane was shot down. Lyle Harris was seven years old by then. When he was little, Lyle would often see a plane in the sky and say, "There goes Daddy." But the man who held him now was a stranger.

"I picked him up and hugged him for a long time, and it didn't bother me that he didn't hug back," Harris remembers. "I knew it would take a little time." Later, as Harris handed out gifts for his family, he saw Lyle standing in a corner, watching. "I turned and opened my arms toward him and he came running, jumped in my lap, and threw his arms around my neck in a big hug. Again, *Oh thank you, Lord*."

— **Carlyle "Smitty" Harris**
U.S. Air Force, Captain, F-105 fighter pilot
Vietnam: 1965–73 (POW)

"K" is the one missing letter—"C" is used instead.

Captain John Parsels and other POWs are welcomed at Gia Lam Airport in Hanoi upon their release from a prisoner of war camp. *Photo courtesy of NARA.*

Catapult officer on board USS *Hancock* (CVA-19) signals pilot of A-4 Skyhawk for launching. *Photo courtesy of NARA.*

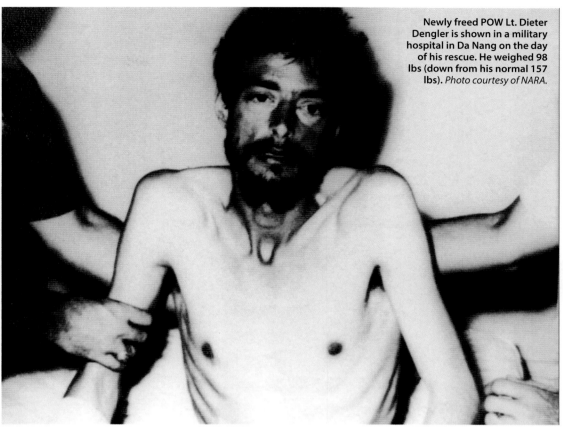

Newly freed POW Lt. Dieter Dengler is shown in a military hospital in Da Nang on the day of his rescue. He weighed 98 lbs (down from his normal 157 lbs). *Photo courtesy of NARA.*

Saluting the Moon

The American prisoners of war in Vietnam didn't know about Neil Armstrong's historic walk on the moon until a year after it happened. But once they found out, they had a little fun with it at their captors' expense.

"They had these horrible interrogations, where they were just trying to give you propaganda," says Edward Martin, a POW at the "Hanoi Hilton" who was captured after his A-4 Skyhawk jet was hit over North Vietnam. During one such interrogation, the North Vietnamese captors boasted that Neil Armstrong didn't need to tell the Vietnamese what the craters of the moon looked like, because they already knew.

"I knew immediately that we had succeeded in landing on the moon, so we started to stand outside and salute the moon," Martin said. "They wanted to know why we were doing this. We said, 'You'd better take care of us and treat us properly, because we know your Tet Festival every year is predicated on the cycles of the moon. And now that we own the moon, we will not allow you to use it for your Tet anymore.'"

In prison, Martin learned to keep his mind and body as active as possible. He created word puzzles in his mind, and then completed them. He went over basic arithmetic, calculus—even worked hard at memorizing poetry long since forgotten. "Anything to keep your mind active is terribly important in a situation like this," Martin recalls.

During his captivity, Martin was severely beaten and tortured, and kept in solitary confinement for 13 months. His captors broke his shoulder and never gave him medical attention for it. "When we left Vietnam," he says, "it was like going from hell to heaven." Martin was among the 591 American POWs returned alive to the United States. His experience as a prisoner, he says, had a profound effect on his life. "First of all, it gave me a joy of living," he explains. "It made me appreciate my family, children, and friends. That experience taught me, more than anything else, the importance of an education."

— Edward Holmes Martin
   U.S. Navy, Captain, fighter-bomber pilot, Attack Squadron 34, USS *Intrepid* (CVS-11)
   Vietnam: 1967–73 (POW)

Martin remained in the Navy, eventually becoming Vice Admiral and Deputy Commander in Chief, U.S. Naval Forces Europe. He died in 2014.

Men of Company F, 2nd Battalion, 4th Marines move along a punji-stake-embedded trench. *Photo courtesy of NARA.*

A Year of Firsts

As part of the 1st Battalion, 3rd Marines, Dan Hall and his fellow Marines made history. They were the first U.S. ground combat battalion to enter combat operations in the Vietnam War, the first Marine unit ever to depart a base outside of the theater of operations and deploy by air into the front in less than 24 hours, and the first Marines to use the new concept of helicopter insertion to strike enemy positions.

Although he was part of all these "firsts," being a Marine wasn't new to Hall. He'd actually enlisted in 1960 and then re-enlisted before being sent over to Vietnam. He and his best friend, Hugh "Sully" Sullivan, were both promoted to sergeant and both assigned to the newly formed 1st Battalion at Camp Pendleton, where they trained to be inserted somewhere in the Pacific.

The battalion was initially assigned as a base defense force at the Da Nang Air Strip in Vietnam, "but that only lasted about 30 days," Hall says. "Then we were out in the bush—the first unit to engage with the enemy." In June 1965, Sully's squad was hit by several rounds of sniper fire outside a village, and Sully was shot. "He'd just been telling me about his daughter being born back home and how anxious he was to see her, and four days later, he was killed," Hall recalls. "It broke my heart."

Only a few days later, Hall made history again as one of four Marines featured on what became an iconic cover of *Life* magazine. Hall first saw the magazine while lying in a hospital bed in Okinawa. Diagnosed with malaria, he'd been medevaced out of Vietnam shortly after the photo was taken.

Although he only served four months in Vietnam, Hall suffered physically and emotionally for more than four decades. But 40 years after his return, he received a phone call that brought him great joy. Roger Warren, a 1st Battalion historian, called to tell him about an upcoming battalion reunion—and mentioned that Sully's daughter, the one who had been born four days before he was killed, would be there.

The two met in an emotional reunion. "She came to my house to visit me," Hall says. "I got to meet the daughter Sully never met."

— Dan Hall
U.S. Marine Corps, Sergeant, 3rd Platoon, Company A, 1st Battalion, 3rd Marines, 3rd Marine Division
Vietnam: 1965

Semper Fi

Dennis Howland says he saw enough blood and death in Vietnam to give him nightmares for years. So he would rather talk about a quieter story, the kind that took place in the villages where Navy corpsmen brought rudimentary medical care and Army engineers built fresh-water wells. As a Marine, Howland often accompanied the corpsmen to provide security, and while he was there he would become enmeshed in village life.

What he remembers most, he says, are the children: how they were always so happy to see him; how they would go off to play and then come back and draw pictures in the dirt to let the Americans know where the Viet Cong were hiding; how one little girl cried when he left for America.

The girl's mother had died, and her brother was conscripted into the South Vietnamese Army. Later, Howland learned, she was taken to an orphanage, so when he returned to the States, Howland set in motion a plan to adopt her. But the orphanage was hit by North Vietnamese rocket fire and the little girl was killed.

All of this—the kindnesses and the deaths—was too hard to explain when he was confronted by anti-war demonstrators after he got home. Like other returning vets, he was spat on and yelled at. In his parents' hometown in Missouri, though, he was invited to sit on the stage during a Veterans Day celebration, next to a group of Gold Star Mothers who had lost sons in the war.

"I promised those mothers," Howland says, "that as long as I breathed, I'd never let the world forget how important their sons were to the history of this country."

Howland has kept his promise. He moved to Utah, where he has worked tirelessly for Vietnam Veterans. He started a chapter of Vietnam Veterans of America and became the Utah organization's president. He lobbied to establish an annual Vietnam Veterans Day. He organized "The Welcome Home Parade They Never Got," and has found space and funding for a permanent replica of the Vietnam Memorial Wall.

"It's about all Vietnam Veterans," Howland says. "Those killed, those injured, those affected by the unseen scars of war, and those still trying to make that journey home, those still waking up from the same nightmares after 50 years. It is about each and every one of them."

One day last fall, a woman rang Howland's doorbell. "You're the guy who's always in the paper, the one bringing the Wall," she said, and began to cry.

She told Howland her story: how the older brother she idolized had been killed in Vietnam when she was in high school, and how she had turned her grief into anger. She marched against the war. She threw garbage at Soldiers.

"I've spent 50 years wondering if I'd ever have the chance to undo what I did," she told Howland. When she saw him on TV and saw him at the parade, she said, "I knew you were the one I could apologize to. I'm asking you, on behalf of Vietnam Veterans, to accept my heartfelt apology."

"Now we were both crying," Howland says.

He understands what it's like to lose people you love. His hometown of Council Bluffs, Iowa, lost 26 of its boys in Vietnam; nine were from Howland's high school. One of his best friends, as well as the friend's brother, plus many Marines that Howland served with, died in Vietnam.

"That's why I do what I do today," he says. "Because I have friends on the Wall. My men are on the Wall. And as all Marines will understand—Semper Fi, my brothers and sisters of the Wall. Semper Fi."

— Dennis Howland
U.S. Marine Corps, Sergeant, 3rd Marine Aircraft Wing
Vietnam: 1966–67

Men of Company H, 2nd Battalion, 7th Marines move along rice paddy dikes in pursuit of Viet Cong.
Photo courtesy of NARA.

Code of Honor

The first time Jerry Coffee was allowed to write a letter home as a prisoner of war, he had to write with his left hand because his right arm was broken and had been left untreated. "I'm being treated well," he wrote to his wife, Bea. And then he criticized the U.S. involvement in the Vietnam War.

The next time he had a chance to write a letter, he wanted to figure out a way to let Bea know the truth. He hoped she would remember something he once told her: "If I'm ever captured, I'll put little dots under certain letters to spell out a secret message."

To spark Bea's memory about the code, he wrote in a letter to their son, "David, I can just see you playing with your little connect-the-dot books to get the big picture." The included secret message read: "Forget B.S. in first letter. This camp in Hanoi. Torture." He also spelled names of other POWs whose families and the U.S. military didn't know were alive.

Eventually he was allowed to read a letter back from his wife, where he got two pieces of good news. Bea had been pregnant at the time of his capture, and now they had a new son. And when he was able to read the handwriting in better lighting, he noticed a tiny dot under a lowercase *i*. He quickly scanned and found more tiny dots.

Bea's coded message read: "Send names and locations. Hang on." He was elated to realize his code message had been received and answered.

But Coffee was never able to use the dot code again. Future letters were written in the presence of a North Vietnamese officer, and were limited to six lines. So it took some creativity to come up with word associations to reveal names of other POWs, without the enemy suspecting.

In one letter to his family, Coffee tried to include the name of a young Navy radar officer, Ensign Dave Rehmann. "Stevie," he wrote, "I often picture you playing with those little space figures you loved. There was Superman, Batman, and Rayman. As I recall, Rayman was your favorite, right?"

Having a grateful attitude—counting his blessings—and writing down his thoughts helped Coffee survive the seven long years of his captivity. His autobiography as a POW, *Beyond Survival*, was published in 1990. "We're all a lot tougher and more capable than we think we are," he says. "Unless we're challenged, we don't know that about ourselves."

— Gerald "Jerry" Coffee
U.S. Navy, Commander, aviator, Reconnaissance Squadron 13, USS *Kitty Hawk* (CVA-63)
Vietnam: 1966–73 (POW)

A Delicate Operation

When Capt. Harold Dinsmore first saw the x-ray, he thought his colleagues were playing a trick on him—because what the picture revealed that October evening in 1966 was an unexploded 60mm mortar round in the chest of a South Vietnamese soldier.

Dinsmore was chief of surgery and the senior surgical officer on duty at the Naval Support Activity Hospital in Da Nang. When the soldier, 22-year-old Nguyen Van Luong, was brought in, Dinsmore recalls, "it was immediately obvious what had to be done." And he was the one who was going to have to do it. "With the gravity of this situation, I felt I could not ask or order anyone else to do the surgery," he says, describing it as one of the most harrowing of his Vietnam memories.

A demolition expert, Engineman 1st Class John Lyons, was called in and gave this assessment: the mortar round in the soldier's chest contained between one and two pounds of TNT that could go off at any time, even without being handled.

The patient was taken to the operating room by stretcher, "and I never saw such careful, tiptoeing stretcher carriers," Dinsmore says. The soldier was placed on the operating table, stretcher and all, and then sedated, intubated, and attached to an automatic respirator. Then the anesthesiologist left the room. "I had decided no one should be there who didn't have to be. Only Lyons and I would stay."

Dinsmore was warned not to twist or move the mortar round at all during the surgery, and to lift it straight from the chest wall. Every second, with the shell just a foot from his face, "I thought my world was going to end."

To make matters worse, the patient's blood-soaked shirt was badly entangled in the mortar round's tail fins. "With Mayo scissors, the heaviest we had, I spent an additional harrowing 10 minutes cutting through multiple folds of heavy, wet cloth to get it free."

When Dinsmore finally lifted the mortar round from the soldier's body, Lyons took it to a nearby sand dune and defused it. He later removed the explosive charge, then gave the empty round to the surgeon as a keepsake. Nguyen survived and eventually returned to full duty. For Dinsmore's courageous actions that day, he later received the Navy Cross, the service's second-highest combat award.

— Dr. Harold "Hal" Dinsmore
U.S. Navy, Captain, surgeon, Naval Support Activity Hospital, Da Nang
Vietnam: 1966–67

Excerpt from: Herman, Jan K. *Navy Medicine in Vietnam: Passage to Freedom to the Fall of Saigon.* The U.S. Navy and the Vietnam War. Eds. Edward J. Marolda and Sandra J. Doyle. Washington, D.C.: Naval History & Heritage Command. 2014

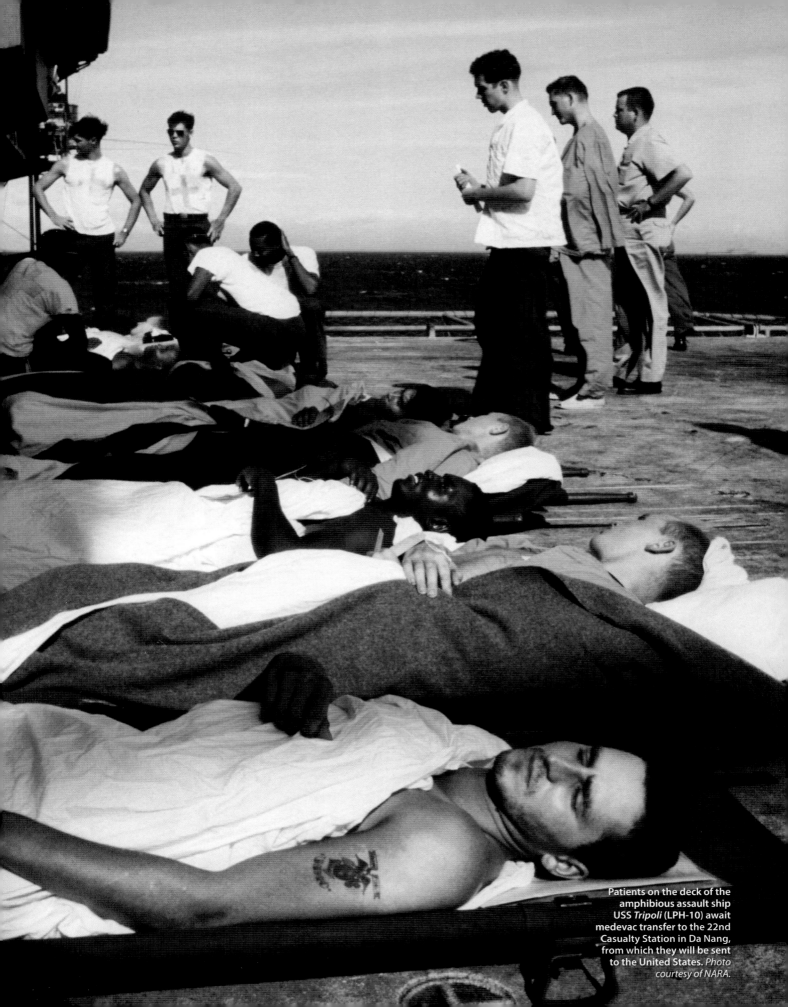

Patients on the deck of the amphibious assault ship USS *Tripoli* (LPH-10) await medevac transfer to the 22nd Casualty Station in Da Nang, from which they will be sent to the United States. *Photo courtesy of NARA.*

Specialist 4 R. Richter, 173rd Airborne Brigade looks forward while Sgt. Daniel E. Spencer stares down at a fallen Soldier. They await a helicopter which will evacuate their comrade. *Photo courtesy of NARA.*

In Memory of Them

On his worst day in Vietnam, Ken Garthee was on a reconnaissance mission near the DMZ, scouting the triple-canopy jungle and the open areas at the base of a mountain called "the Rock Pile." With him were 50 Marines, including a new Scout Sniper named Jack, a quiet guy, newly married.

They had just forded a river. Garthee and Jack moved a short way up the ridge to get a better view of enemy movement, and Garthee took off his flak jacket and climbed a tree to check their location. Looking through his binoculars, he saw an enemy soldier staring back at him, a scout for NVA soldiers holed up in the Rock Pile.

As soon as Garthee jumped down from the tree, bullets started flying. He tried to dive for cover but was hit in the chest. Jack ran to his side and shielded him with his body. Soon the rest of the company was running up the side of the mountain, shooting and screaming. The forward observer called for air support and a helicopter to evacuate the wounded.

Garthee had an exit wound the size of a bowling ball, and his veins and left lung had collapsed. A Navy corpsman placed a cellophane wrapper from a pack of cigarettes onto the entrance of his chest wound and packed the exit wound with gauze. And then Garthee felt himself floating up from his body. "I remember every moment," he says. "It was calm. And there was a hand on my shoulder. There were never any words spoken—it was more of a feeling of reassurance." When the battle subsided, he says, he felt a gentle nudge, then he felt himself sliding back down into his body.

He was put into a body bag, and the bag was zipped up. A KIA tag was attached to the zipper. He was carried to a loading area, and as the body bag was dropped onto the ground, he let out a groan. "Hey, Garthee's not dead!" Jack shouted. They unzipped the bag and gave him water, and put him on board the medevac chopper. The body bag was left behind, still with the KIA tag attached to the zipper.

The next day, Marines told his parents that their son had been killed in action—but his mother refused to believe it. Garthee later learned that after he was medavaced out, another air strike was called in. One of the six bombs that dropped ricocheted off the side of the Rock Pile, killing the 16 Marines left behind, as well as the four Marines who had carried him to safety. One of them was Jack.

"Every time I help a family in need today," says Garthee, "I do it in memory of and in honor of the guys who threw me on the helicopter that day. Everything I do, I do in memory of them. Every day."

— **Ken Garthee**
    U.S. Marine Corps, Private First Class, infantry, 3rd Battalion, 3rd Marines, 3rd Marine Division
    Vietnam: 1966–67

Explosive Ordnance Disposal (EOD) team makes landfall on a muddy riverbank. *Official U.S. Navy photo courtesy of Russell A. Elder.*

Love and War

Coming from a small town of just 400, nurse Jeanne Urbin dreamed of working in a big Chicago hospital. But when her mother insisted the city was too dangerous, Jeanne signed up for the Army instead. While stationed in Fort Carson, Colorado, she met and became engaged to Army officer Brian Markle. She told him if he got orders for Vietnam, she would cancel the wedding. "So I waited until the night before the wedding to tell her," Brian admits.

Four months later, in December 1966, the newlyweds arrived together in Vietnam. Brian was a medical logistics officer, handling everything from ordering supplies to maintaining vehicles. Jeanne worked at the 24th Evacuation Hospital in Long Binh, where they saw 30-35 casualties a day. At times, surgeons were operating 72 hours straight; at night, she was the only nurse overseeing a 33-bed unit.

They had unique experiences, Jeanne recalls. The hospital ran out of bandages one week, and they had to use the *Stars and Stripes* newspapers as dressings until they got more sent from the U.S. or the depot. When a patient's death was imminent, Jeanne would often sit and hold the Soldier's hand so he wouldn't have to die alone. Once, a 19-year-old patient who had lost both his legs asked Jeanne to write a letter to his wife, to find out if she would welcome him home.

The staff learned to improvise. They had no running water, so "every morning at 4, the sergeant or corpsman would go out and set fire to a great big barrel of water and let it heat until 6," Jeanne said. From that barrel, the nurses filled pitchers of water to bathe their patients.

Jeanne recalled the camaraderie of the Soldier-patients. "If it hadn't been for them, we couldn't have gotten the job done," she said. "If there was someone in the next bed that had one arm that he could use, he got up and bathed the boy next to him. If he was in a wheelchair, he would go help anywhere. These were 17, 18, 19, 20-year-old boys and they were just as scared as I was, and with devastating injuries. But it was wonderful how they took care of each other."

— **Brian Markle**
  U.S. Army, 1st Lieutenant, medical logistics officer, 58th Medical Battalion
  Vietnam: 1966–67, 1970–71

— **Jeanne Markle**
  U.S. Army, 1st Lieutenant, nurse, 24th Evacuation Hospital
  Vietnam: 1966–67

Brian, who retired as Lieutenant Colonel, used his medical logistics experience in Saudi Arabia during Desert Storm in 1991.
Jeanne passed away in 2009 from pulmonary fibrosis; her death certificate states it was due to Agent Orange.

Watching the War from on Deck

As a Navy doctor on the hospital ship USS *Repose* (AH-16), Joel Johnson saw the war from a different perspective than most who served in Vietnam—two or three miles off the coast of South Vietnam, near the cities of Hue and Dong Ha. "We were hardly ever out of sight of land," says Johnson, who at night would sometimes stand on deck and watch the war. "You could see the explosions going on, and one evening we got a close look at the battleship USS *New Jersey* (BB-62) as it cut loose with its 16-inch guns."

Helicopters had improved so much by the Vietnam War that they could fly directly from a battlefield to a hospital, so it was easier and safer to reactivate and use an existing hospital ship like the *Repose*, which could be located just off the coast. "So basically what I did was like Hawkeye [on the TV show *M*A*S*H*], only our hospital setting was a little bit better than the ones on land, a little more permanent," Johnson explains.

But the better setting and faster helicopters didn't make it any easier emotionally for the medical personnel who treated wounded servicemen. More than 40 years later, Johnson still gets emotional when he talks about his patients. One man, perhaps the most serious injury Johnson treated, lost both legs and an arm; another patient required 65 units of blood, six or seven times more blood than normally flows through a body.

"The worst time for being busy was during the Tet Offensive in January 1968," he recalls. "Our ship was supposed to have 200 beds on it for patients—but we had 700 patients, so the lesser-injured slept on the deck."

But some good came out of his experience, Johnson says. They had a few Soldiers come aboard with gas gangrene, a condition where life-threatening organisms live in dead tissue. If there was blood circulation near the wound, "antibiotics could kill the organism," he explains, "but antibiotics can't get to it because dead tissue doesn't have any circulation."

The ship had a decompression chamber,* and out of desperation, Johnson's team put a dying patient into the chamber. "By golly, it worked," he says. "It didn't take long for word to spread through all the units, so I ended up taking care of quite a few cases of gas gangrene."

— Joel Johnson
U.S. Navy, Lieutenant Commander, physician, USS *Repose* (AH-16)
South China Sea: 1967–68

*A decompression chamber, often used to prevent the "bends" in deep underwater divers, forces more oxygen into the blood, thus strengthening the blood's healing properties.

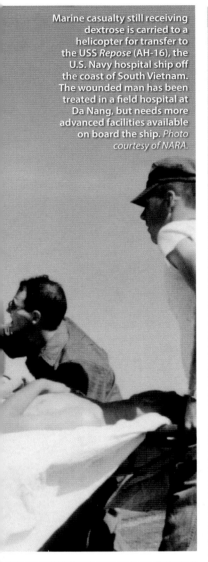

Marine casualty still receiving dextrose is carried to a helicopter for transfer to the USS *Repose* (AH-16), the U.S. Navy hospital ship off the coast of South Vietnam. The wounded man has been treated in a field hospital at Da Nang, but needs more advanced facilities available on board the ship. *Photo courtesy of NARA.*

A Covert Snake Operation

In 1964, while stationed at England Air Force Base in Louisiana, Oak Norton, an Intelligence Officer, received an urgent call from U.S. Air Force Logistics Command. American forces in Thailand "suspected a possible attack from Laos," he explains, "so we were told we had 20 minutes to prepare to board a jet to Thailand." Less than 24 hours later, the Special Ops intelligence staff landed at the Takhli Royal Thai Air Force Base.

Explaining the U.S. Air Force presence at Takhli, Norton says the Thais feared the civil war in Laos would spread into their country, bringing Communist forces with it. Thus the U.S. was allowed to use five bases, in exchange for the air defense of Thailand. As the Vietnam War expanded, U.S. pilots flew attack sorties from Thailand into Vietnam—as well as secret combat missions into Laos, Norton says. Those missions are today referred to as "the Secret War."

Norton's responsibility was to brief U.S. pilots on all intelligence matters. Because he was involved in such clandestine operations, Norton often had some unusual covers—including passing himself off as an expert herpetologist.* "When we arrived at Takhli," he explains, "we were told to billet [live] at some old Japanese barracks left over from World War II. We cleaned out the snakes, bats, and scorpions to make the area livable."

Although most of the abundant varieties of snakes were poisonous, Norton wasn't worried. "I had some knowledge of snakes," he says, "so I felt fairly safe." Interested in herpetology, he traveled to a "snake house" in Bangkok under an assumed identity. "The curator, Mr. Y-Siah, was interested in getting some Kraits—one of the deadliest snakes in the world," Norton recalls. "Poisonous snakes like Kraits and Cobras were responsible for killing thousands of people each year. I told him I was interested in helping him and would bring as many as I could to him."

The undercover identity was a perfect fit for Norton, allowing him to dabble in his hobby while gathering intelligence from Communist sympathizers. Catching snakes also ensured he got his own room. "One night I caught a poisonous snake under the [barracks'] fridge and put him in a big can," Norton recalls. "The next morning when I woke up, my roommate was there, raised up on his elbow smoking a cigarette. I slid the lid off the can to check on the snake. When my roommate saw it, he fell over backwards and ran out of the room in only his undershorts. After that, I never had any more roommates—and my room was smoke-free."

Norton soon had a large collection of poisonous snakes to take to the Snake House in Bangkok—and some good stories to tell his grandkids.

— Oak Norton
U.S. Air Force, Master Sergeant, Air Force Special Operations Forces (1964–65); 109th Military Intelligence Battalion, Air Force Logistics Command (1965–67) Thailand: 1964–67

One who studies reptiles and amphibians.

U.S. Army troops on the move in South Vietnam. Hovering in the background are Army CH-47 Chinook helicopters. *Photo courtesy of NARA.*

Oak Norton is on the far right.

On A Wing and A Prayer

The mountains were shrouded with dense clouds the day Bernard "Bernie" Fisher and Dafford "Jump" Myers were part of a mission to strafe enemy positions in South Vietnam's A Shau Valley. It was March 10, 1966, and the pilots were trying to help a U.S. Special Forces camp under attack by the NVA.

Fisher and Myers were flying Douglas A-1E Skyraiders low enough to see their targets—which meant they were also within range of enemy fire. When Myers' plane was hit, he was too low to bail out, and crash-landed on the camp's runway.

"I called the command post and said we have a pilot down and he was probably killed in the crash," Fisher remembers. But then he saw Myers scramble away from the wreckage and dive for a ditch. Fisher called for a rescue helicopter, but when the promised chopper didn't show up, he knew he needed to act quickly.

"Enemy soldiers were everywhere. I realized we had to get Myers out of there or he wouldn't make it," Fisher says. While other A-1E pilots flew cover for him, Fisher brought his Skyraider through the smoke and fire and touched down, then skidded toward the end of the runway—and realized it was too short.

"I ran out of runway and brakes," he explains. "I damaged the right wing, but got [the plane] turned around and taxied back down the runway." He was taking enemy fire when he saw Myers running to the plane. Then he lost sight of him—until there Myers was, crawling up on the wing of the plane. "I jumped up and pulled him in headfirst," Fisher recalls. "We didn't even strap in. We took off and I gave it all the throttle she had." Fisher took off under fire and headed for home base.

When Fisher and Myers landed back at Pleiku, the ground crew found 19 bullet holes in his plane. The next year, Fisher was awarded the Medal of Honor by President Lyndon Johnson at the White House. Never forgetting his rescue, Myers called Fisher every year on March 10. After Myers died in 1992, his daughter kept up the anniversary tradition for 22 more years, until Fisher's death in 2014.

— **Bernard Fisher**
   U.S. Air Force, Major, pilot, 1st Air Commando Squadron
   Vietnam: 1965–66

Major Fisher became the first living Air Force Medal of Honor recipient. He retired as Colonel in 1974.

Men of Battery C, 2nd Battalion, 32nd Artillery load a round into a 175mm howitzer. *Photo courtesy of NARA.*

One for the Road

Bill Linderman and 21 other new Soldiers had just finished ordnance training at Fort Dix, New Jersey, and reported to McGuire Air Force Base with orders for Vietnam. "Someone brought in some booze," he recalls, "and so we partied that night." At 0400 the next morning, Linderman and his fellow Soldiers joined 140 recent graduates from the Army's parachute jump school, and all boarded their plane. The chartered TWA flight was bound for Vietnam via Alaska, and the airborne guys provided the in-flight entertainment with their boisterous singing.

In Anchorage, the men made a beeline for the nearest bar, where before long a party was well underway. A man in a civilian flight jacket, sitting at the bar with his wife, joined in the fun. "Pretty soon he started buying drinks for the airborne guys, and then he tried on one of their jackets and a hat," Linderman says.

After a while, the man drunkenly declared he was going with them to Vietnam. The paratroopers started ribbing him that he "didn't have the guts" to go. The man insisted he was going, so as a joke, when they were recalled to the plane, the Soldiers snuck him on board wearing the Army jacket and cap. The Soldier who had loaned the jacket went into the plane's lavatory, and the civilian took a vacant seat.

Everyone figured the joke wouldn't last long. But the plane was running late after the delay, and the crew was in a hurry, so no one checked the passenger manifest until it was too late. The plane—which coincidentally had the flight designation of "Whiskey 5"—quickly took off for Japan. About 20 minutes into the flight, the stewardess went down the aisle counting passengers and stopped at the extra "guest," who by now had removed the military jacket and hat and was sitting nonchalantly in a big red flannel shirt, drinking from his pint. When asked to identify himself, the man bellowed, "My name is Johnson, and I'm going with them!" The passengers roared with laughter.

The stewardess, however, was not amused. When the plane landed in Japan, the Air Police was waiting, and a now sober and compliant Mr. Johnson no longer wanted to "go with the boys." The police escorted him away. "By the way," they told him, "your wife is still waiting for you back at that bar."

— **Bill Linderman**
   U.S. Army, Specialist 4, wheeled-vehicle mechanic, 148th Ordnance Company
   Vietnam: 1966–67

His Last Day in Vietnam

Captain James Drake James was within hours of returning to the U.S. after his first tour of duty in Vietnam. He'd already turned command of his company over to his replacement and was waiting to board the plane when the call came in—one of the company's reconnaissance teams was in contact with the enemy.

"The new company commander wasn't around," recalls James, who didn't think twice about what to do next. "I got a helicopter and flew out to where the team was." James and his crew found the team and landed amid a shower of bullets. "We took some hits in the helicopter, but we were able to rescue the team and get them out from the enemy force that had them surrounded," he says. "It gives me goosebumps even now when I talk about it."

During his time in Vietnam, James received several medals, but he never got a medal for what happened his last day. "Nothing was ever submitted," he says. "I just went home. And that's okay. I tell that story because there were a lot of people who did the same sort of thing and never got any awards. A lot of other people who are your true heroes are walking around with no medals to show that."

— **James Drake James**
   U.S. Army, Captain, Long-Range Patrol Detachment, 1st Cavalry Division (Airmobile)
   Vietnam: 1966, 1968

A U.S. Navy in-shore patrol craft PCF-38 of Coastal Division 11 navigates the Cai Ngay Canal.
Photo courtesy of NARA.

1967
The Buildup

In Vietnam, 1967 was "the year of the Allied offensive." Back home, anti-war demonstrators were chanting "make love, not war," while pro-war Americans carried banners urging people to "support the troops." It was a clash of slogans and cultures that embodied the hawk-and-dove political battle that would rage for the next five years.

The year 1967 saw the arrival of the M-16 rifle and the addition of the Cobra gunship to a war that relied more and more on helicopters to hunt down the enemy as well as transport troops. Under the direction of MACV commander Gen. William Westmoreland, the U.S. military increased its large-unit search-and-destroy missions, as well as attacks on North Vietnamese MiG bases and the Ho Chi Minh Trail.

Westmoreland's strategy of attrition aimed to inflict heavy losses on North Vietnamese and Viet Cong forces using superior U.S. firepower. Less visible to the media and public, U.S. advisors at the unit, province, and district level worked to build the capability of the South Vietnamese. Buzzwords that year were *body count*, *Vietnamization*, and *hamlet evaluation*, as the Department of Defense sought to establish mathematical measures of success. Westmoreland predicted victory by the end of the year.

UH-1D helicopters airlift members of the 2nd Battalion, 14th Infantry, 25th Infantry Division to a new staging area during a mission. *Photo courtesy of NARA.*

TIMELINE 1967

January 1967
Number of U.S. in-country troops approaches 500,000.

January 2, 1967
Ronald Reagan sworn in as governor of California.

January 8, 1967
U.S. forces launch Operation Cedar Falls to eradicate Iron Triangle area around Saigon of Viet Cong bunkers and tunnels.

March 1967
U.S. aid to South Vietnam increases by $150 million, bringing the total for the year to $700 million.

April 15, 1967
400,000 in New York and 100,000 in San Francisco protest the Vietnam War.

June 1, 1967
The Beatles release *Sgt. Pepper's Lonely Hearts Club Band*, considered the first concept album, with all songs unified by a common theme.

September 1967
Secretary of Defense Robert McNamara calls for a barrier with fencing and ground sensors to block communist infiltration of the eastern DMZ.

November 1967
General Westmoreland tells U.S. newsmen, "I am absolutely certain that whereas in 1965 the enemy was winning, today he is certainly losing."

November 30, 1967
U.S. casualties in Vietnam reach 15,000.

December 31, 1967
486,600 U.S. troops in Vietnam.

You Do Your Job

At the medical center at Dong Ha that July day, Roger Cox sat with hundreds of other wounded Marines. He'd been operated on the day before at Camp Carroll, but because there were so many wounded and not enough supplies, he'd had the ankle surgery without general anesthesia. Now he was waiting to have his bandage changed, sitting with men who had just been brought in from a fierce battle with North Vietnamese troops near Con Thien. It was the first day of Operation Buffalo.

As the men waited for treatment, a gunnery sergeant ran in to announce that Company B (Bravo), 1st Battalion, 9th Marines had been overrun, and Bravo's 3rd Platoon had been cut off from the rest of the company. It turned out to be the single worst day for Marines in the Vietnam War, with 84 dead and 190 wounded. If you're at all able, the sergeant pleaded, come help us find the missing Soldiers.

Suddenly, wounded Soldiers were standing up, taking off their medevac tags and leaving for the staging area, Cox recalls. This was their opportunity "to experience Semper Fi." Cox had only one boot—the other had been cut off before surgery. So he went to Graves Registration,* where he knew he would find a stack of clothes that had belonged to fallen Marines. He sorted through the pile until he found a boot that would fit over his bulky bandage.

Cox and 250 other Marines were transported to Con Thien, near the DMZ. As they swept northward on foot, under scattered artillery fire, they came upon dozens of bodies of dead Marines and North Vietnamese, all blackened after three days in the 135-degree sun.

"No Marine that saw the results of the NVA on this day would ever consider giving up," Cox wrote in the compilation *My Vietnam: Montana Veterans' Stories Straight From the Heart*. "Your biggest fear was not dying," he adds today, "but not performing in the manner in which Marines are expected to. You do your job."

— **Roger Cox**
  U.S. Marines, Corporal, 1st Battalion, 9th Marines,
  3rd Marine Division
  Vietnam: 1967–68

The unit charged with receiving, identifying, and caring for the bodies of those killed.

U.S. Soldiers hug the earth as another company sweeps toward them, pushing Viet Cong troops in their direction. *Photo courtesy of NARA.*

Playing Shenandoah

One day in the summer of 1967, Sammy Davis's mother went to see the nice folks at the Red Cross in Martinsville, Indiana. She was worried sick, she told them, because her son was serving in Vietnam, and she hadn't heard from him in 63 days. The Red Cross turned to its contacts at the Pentagon for assistance.

Early one morning soon after in the Mekong Delta, as Sammy Davis stood guard out on the unit perimeter, his commanding officer approached him with a stern look. "He walked up to my foxhole and said, 'Private Davis, why haven't you been writing home to your mother?'" Davis recalls. "I told him I wasn't comfortable writing home to Mom about what was happening." Besides, he told his CO, it's so wet out here, the paper turns into spitballs.

"I don't care what you write about," said the CO, "or what you write on. But you will write your mother every day." So Davis did. He wrote about elephants, monkeys, and ducks. "Things that were true," he explains, "but not about the war."

After several weeks of daily letters to his mom, Davis got a package from home. He was hoping it was his mom's famous oatmeal–raisin cookies, but instead it was a small Horner Echo harmonica. On the outside of the box she had written, "I hope this will help you from being so bored."

Davis continued to refrain from writing about what the war was really like. He didn't write home about the night in November 1967, at Firebase Cudgel near Cai Lay, when he was knocked unconscious from an enemy RPG, or how on regaining consciousness, and despite wounds to his back and legs, he grabbed a machine gun to mow down attacking enemy soldiers. He didn't write about how, when he ran out of ammo for his weapon, he was able to repeatedly load and fire anti-personnel shells from a smoldering howitzer. Or how despite multiple injuries, he paddled an air mattress across a canal, with enemy bullets pelting the water around him, to help rescue wounded Soldiers on the other side.

Before all this happened, Davis had taught himself to play that harmonica. Sgt. John Dunlap was the one who encouraged him, the very first night it arrived. "Play Shenandoah," he said, and after several weeks it sounded less like honking and more like music. It turned out that after his first tour in Vietnam, Dunlap had enrolled at a college back home in Maryland, a liberal college full of anti-war protestors. Instead of arguing with them, Dunlap would drive out to the Shenandoah River and just sit there until he felt peaceful again.

Dunlap was killed in Vietnam not long after Davis returned home. A dozen years later, on the night before the official dedication of the Vietnam Veterans Memorial, Davis arrived in Washington, D.C., and went straight to the Wall, even though by then it was 3 a.m. He found Dunlap's name, touched it, pulled out his harmonica, and played.

— **Sammy L. Davis**
   U.S. Army, Private First Class, artillery, Battery C, 2nd Battalion, 4th Artillery,
   9th Infantry Division
   Vietnam: 1967–68

In 1968, Davis received the Medal of Honor for his actions at Firebase Cudgel. He remained in the Army until 1984. The news footage of Davis receiving the Medal of Honor was used in the movie "Forrest Gump."

Rotting Uniforms

The first trip up Tiger Mountain, it rained so hard the enemy soldiers didn't hear Lt. Robert Alekna and his men approach their position. Alekna's point men spotted the enemy under the cover of their ponchos and eliminated them with their M-16 rifles. After a brief firefight, the Cavalry troops were ordered to withdraw so that air and artillery fire could be brought in. After a second firefight, Sgt. Ross Wood, one of Alekna's squad leaders, went missing.

"We knew he was shot, but we couldn't find him," Alekna says. "And it was getting dark out and I was getting orders to get off the hill. Everybody else was off that hill except my platoon, and I said I wouldn't leave until we found him." They finally found him, and Alekna's men somberly carried Wood's body down the mountain.

The next morning, the division and battalion commanders came out to see how Company A was doing—and then gave orders to go back up the hill. When Alekna offered to lead the way again, the division commander asked if the men needed anything. "We could use some new uniforms," offered Platoon Sgt. James Campbell.

"Typically when you're out in the brush, you're out for three, four weeks at a time, and you're wearing the same clothes," Alekna explains. "You don't bathe. You're just grubbing it for a month, and a lot of our uniforms, because you're crawling around in the brush, they're torn. You're perspiring all the time, so the sweat on your body turns to salt, and the uniform begins to rot."

Just to make sure the commander understood, Campbell pointed to Private Van Weeks, who turned around to reveal that the back of his pants was split from the crotch to his belt. "His butt is hanging out," the sergeant said. "No underwear." On their third trip up the mountain, Company A found the enemy had left the area. Two days later, 2nd Platoon got new uniforms.

— **Robert Alekna**
U.S. Army, Lieutenant, platoon leader, Company A, 1st Battalion, 5th Cavalry, 1st Cavalry Division (Airmobile)
Vietnam: 1967–68

Firing M79 grenade toward a Viet Cong position during an operation near Da Nang. *Photo courtesy of NARA.*

Born-Again American

After nearly six years as a prisoner of war in North Vietnam, Jay C. Hess came home a changed man. "I went to Vietnam as a fighter pilot," he says. "I came back a 'born-again' American."

Three weeks after arriving in Cam Rahn Bay in 1967, Hess received orders diverting him to Takhli Air Base in Thailand. He would be flying F-105 Thunderchief missions over North Vietnam, "which at that time was one of the most highly defended air spaces in the world or in history," Hess says.

His thirty-third mission over North Vietnam in August 1967 was his last. After he released his six 750-pound bombs, Hess's plane was hit hard. He knew he had to get out of there fast.

"So I lit the afterburner, which I shouldn't have done," he recalls. "Suddenly the cockpit filled with fire and the aircraft started down, doing an outside loop. I pulled up the ejection handles and about that time heard over my radio, 'Shark 4 is torching.' I later learned the air rescue personnel searched for me for three hours and finally concluded that I was dead." When Hess came to, he was face down in the dirt. Locals captured him, interrogated him in a cave, and took him to the "Hanoi Hilton." Despite more intense questioning there, Hess refused to give basic information.

"They knocked me to the floor, tied my hands behind my back, rotated them up over my head and tied them to my feet in front," Hess says. "I was sweating something awful, and the pain reached a point where it just didn't hurt any worse. Hours later, I was untied, but my hands were paralyzed. Not wanting to go through that again, I said, 'OK, I can tell you what kind of a plane it was.'"

After two weeks of torture, Hess was put into solitary confinement for two months. His hands began to heal. "It's almost a spiritual experience, because I'd never felt so helpless and dependent," Hess says of his years as a POW. "There's a lot of praying, a lot of reflecting back and thinking about your life, and a lot of regrets. I dreaded the thought of the Air Force car pulling up to my house and telling my wife and five children that I was missing—or worse, dead."

Over the coming years, Hess was given cellmates, and the prisoners were moved several times. Hess was permitted to write his first letter in two and a half years to tell his family he was alive. In February 1973, a group of prisoners were told they were going home. In March, Hess heard he would be in the next group to leave. "We were skeptical, but the next morning, the gates opened up and these buses came in to take us to the airport," Hess recalls. "Along the road, we could see the tail of this airplane, a C-141, and it had an American flag painted on the tail. And everybody's poking everybody else: 'Hey, look at that flag!' Seeing that red, white, and blue—I mean, it is outstanding because of its color, but also because of what's changed in you, and your feelings about your country."

When his plane landed in California, his wife and five children were there to greet him with squeals and tight hugs. Imagining that moment had kept Hess going for the 2,029 days of his captivity.

— Jay C. Hess
U.S. Air Force, Captain, pilot, 357th Fighter Squadron, 355th Tactical Fighter Wing
Vietnam: 1967–73 (POW)

An F-105 drops bombs on a target in North Vietnam. *Photo courtesy of NARA.*

Living on RC Cola

Bernadette "Bernie" Sanner lived on cases and cases of RC Cola during her year as an Air Force nurse at Cam Ranh Bay, South Vietnam. "I seriously didn't eat," she says. "I bought a case every other day, and that's primarily what I lived on." And it wasn't because the food was bad—although it wasn't great. Sanner survived on RC Cola simply because she was too busy. "You had to go to the dining room or cafeteria to eat," she explains.

For the entire time Bernie was in South Vietnam, she and two corpsmen were in charge of a unit that typically had at least 80 wounded Soldiers. "We worked 12 hours a day, 7 days a week, with an occasional day or two off to help you forget about it," she recalls. "And we did everything—we were pretty independent."

Doctors came in twice a week to make rounds, Sanner remembers, but she wrote the orders, made medication changes, determined transfers, and shouldered other responsibilities. "The days were all the same," she continues. "There wasn't a difference between Tuesdays and Saturdays—all the days ran together."

Sanner had been out of nursing school for barely two years when she was sent to Vietnam, arriving on the Fourth of July and landing amid an enemy mortar attack. "It was horrific," says Sanner, who was required to wear a flak jacket and helmet on duty. "Everybody told me that Cam Ranh was the safer of the bases, but we were mortared a lot while I was there."

During mortar attacks, guards were stationed at the hospital doors in case the enemy penetrated the base, but Sanner took care of the patients, covering those who couldn't be moved with mattresses and whatever else they could find for protection.

Too exhausted to eat at the end of her shift one day, Sanner collapsed in her bottom bunk, and slept so soundly that almost nothing woke her—not even the rats. "We had a litter of them, and I guess one night they came out and were crawling up my arm," Sanner says. "My roommate on the top bunk was able to get out, and she went and got the exterminator. I woke up and he was exterminating our place, but I never felt a thing. That's how hard I slept."

— **Bernadette "Bernie" Sanner**
  U.S. Air Force, 1st Lieutenant, nurse, 12th USAF Hospital
  Vietnam: 1969–70

The fourth WAF (Women in the Air Force) officer assigned to Vietnam (lower left) and the first five enlisted women in WAF arrive at Tan Son Nhut Air Base. *Photo courtesy of NARA.*

Members of the first U.S. Air Force Combat Control Team watch paratroopers drop into a Viet Cong-infiltrated area near Saigon. *Photo courtesy of NARA.*

It Was Not the Right Thing to Do

Many American Soldiers in Vietnam had a saying: "We rule the day; but the enemy rules the night." Over 40 years later, the war still deeply affects Lee Alley's nights. The former 1st Lieutenant has trouble sleeping, and his dreams are filled with self-doubt and regret.

"I can't stop analyzing some of the decisions I made as a commander," said Alley, who shared some of his experiences in his book, *Back From War*. For his actions during the war, he received the Army's Distinguished Service Cross, Silver Star, Bronze Star, and two Purple Hearts. But he hid those awards for years, as they were reminders that not all his men returned home alive.

"You're 21 years old and have command of 200 men," he said, "and you must make decisions in an instant, based only on a radio and communication. You can't judge yourself on the information available today; you can only judge yourself on the information you had back then."

One day during the war, Alley was stopped from making a terrible decision, thanks to the bravery of a young South Vietnamese interpreter. A call came over the radio to send a couple of tracks (armored personnel carriers) to check out a possible Viet Cong sighting. Alley decided to send Lt. Dick Bahr, a close friend and his most experienced platoon leader. A few minutes after Lt. Bahr's tracks left the compound, a tremendous explosion rocked the earth. Alley and his men rushed to the scene to see a smoke-filled crater, where Bahr's M113 track and its occupants had been blown to pieces.

"The horror sent my mind reeling and instantly made me sick," Alley said. Using rain ponchos as makeshift stretchers, they began the gruesome task of collecting body parts.

A search squad found the young Viet Cong soldier who likely detonated the bomb. As they brought the bound prisoner to him, Alley began to shake. "My mind suddenly is not my own," he recalled. "All I can see is the carnage, Dick Bahr's and the others' arms, legs, heads. All I can smell is dirt, mud, gunpowder, blood and the odor of death of my men. I get a strange roaring inside my head. Rage and a vengeance-seeking insanity overcame me."

Alley drew his razor-sharp survival knife. The prisoner dropped to his knees and began to plead for his life as a rage-filled Alley screamed that he was going to give him a slow, miserable death for killing his men. "Your ears first, then your eyes!" Alley shouted furiously, grabbing the man's left ear and slicing through the cartilage.

A shout broke through Alley's fog of rage. Sgt. Trang, Alley's trusted South Vietnamese Army interpreter, grabbed his arm. "No! You cannot do this! You are too good a man," Sgt. Trang pleaded. "I owe you my life and I now beg you to spare this one. If you slaughter him, it will trouble you forever. I know this. Please, *Trung Uy*,* do not degrade yourself like this in front of us."

Alley turned and looked into Trang's eyes, and then down at his own bloody, shaking hands. *What have I become,* he thought in despair. He let the knife drop to the ground and sent the prisoner away.

Later that night, in his dimly lit tent, tears blurred Alley's vision as he wrote condolence letters to the families of the 26 men he lost that day. He could see the faces of every man. "They were good buddies," he reflected, "and finer men never walked this earth."

Looking back on his interpreter's interference, Alley said, "I really respected that young man, and when I looked into his eyes, I realized [killing] was not the right thing for me to do. I'm absolutely grateful that he stopped me, because when I look back at all the mistakes in my life, that is one thing I don't have to live with. I can honestly say that in my heart, I never crossed the line. I tried to do the right thing."

At a reunion of Alley's comrades more than 45 years later, he recalled the experience with three of the men who had helped him pick up the bodies. "It was one of the most difficult things we ever experienced," he said. "But as hard as it was, we are at peace with how we handled it."

— **Lee B. Alley**
   U.S. Army, 1st Lieutenant, reconnaissance platoon leader,
   5th Battalion, 60th Infantry, 9th Infantry Division
   Vietnam: 1967–68

   "Trung Uy" is the Vietnamese title for the rank of 1st Lieutenant.

Finding a Father

In 1995, Lori Goss-Reaves placed an ad in a military magazine seeking anyone who had served in the Vietnam War with her corpsman father, Larry Jo Goss. He'd gone missing in action on Valentine's Day 1968. Although his remains were officially identified and returned to the U.S., his dog tags were never found. For 27 years, she'd been hoping he might still be alive.

When Dr. Jerry Behrens saw the ad, he immediately recognized Goss's name. As a battalion surgeon during the war in Vietnam, Behrens met the young corpsman in December 1967.

"I wondered if I really wanted to contact her," says Behrens. "Because some of it was horrific." After thinking it over for a couple of days, he decided he would. "All I could do was be honest, and let her know that her father meant something to me."

For Goss-Reaves, the doctor's kindness was immeasurable. "I will never forget Dr. Behrens so gently telling me that he knew without a doubt that my father died in Vietnam, because he was the one who identified his remains," she says. "He let me know that my dad really was someone special, who made an impression on him."

Portraits of Dr. Jerry Behrens (top) and Larry Jo Goss by artist Eric Reaves

Behrens remembered that Goss's goal had been to become a doctor, and that he had worked and saved money for college, and was accepted at Ball State University. Thinking he had a college deferment, he was shocked to receive a draft notice. He was dismayed to learn his mother had spent his admissions deposit instead of sending it to Ball State. It was a story that Goss-Reaves already knew, and she was touched that Behrens had remembered it all these years later.

"It exemplifies the type of man Dr. Behrens is, trustworthy and willing to listen to a young corpsman's story of heartbreak," she says.

Goss died February 14, 1968, on a steep ridge near the Ca Lu Combat Base, later known as Valentine's Ridge. "There was a very bloody fight, with 10 Marines killed, plus Larry. He was hit by mortars and killed on the spot," says Behrens.

Due to fog, rain, knee-deep mud, and the siege of the nearby Khe Sanh combat base, the bodies weren't recovered for over three weeks. "Every day we'd look out on that ridge, knowing they were out there, but we had orders not to go."

Finally the decomposing remains of the Marines were carried down the steep, muddy ridge and laid out on the ground for Behrens, the chaplain, and the battalion's executive officer to identify. Most were identified by their dog tags, but Goss had his wallet in his breast pocket.

Goss-Reaves was six months old when her father died. Although her childhood hopes of finding out he was still alive were crushed, she was grateful to hear about the father she had never known. Behrens also invited her to a Marine Corps reunion in Orlando, Florida.

"And those Marines just took her in," he says. "She was able to find out what kind of guy her dad was from the men who lived and fought with him."

Knowing her father had dreamed of becoming a physician motivated Goss-Reaves to earn a doctorate degree in social work—earning the title "Doctor Goss-Reaves." She was the first in her family to go to college, 18 years after her father was killed. The oldest two of her four sons have graduated from college, and her daughter is a college freshman. "My dad would be so proud of them," Goss-Reaves says.

Behrens became an orthopedic surgeon after the war, and now volunteers with Semper Fi Odyssey, a group that provides help for wounded, injured, or ill Marines. "Those shared experiences among Marines and their corpsmen and doctors," he says, "bind us together in something that transcends time."

— **Jerome "Jerry" Behrens**
U.S. Navy, Lieutenant, battalion surgeon, 3rd Battalion, 9th Marines, 3rd Marine Division
Vietnam: 1967–68

— **Larry Jo Goss**
U.S. Navy, Hospital Corpsman 2nd Class, 3rd Battalion, 9th Marines
Vietnam: 1967–68

Goss-Reaves is currently writing a book, "Finding More Than My Father: A Journey Into the Heart of a Veteran," about the search for her father and the relationships that developed with his comrades.

Marines take cover during the Battle of Nano Village, when Viet Cong attacked during Tet Offensive. *Photo courtesy of NARA.*

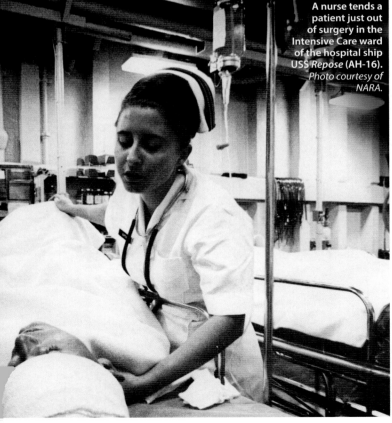

A nurse tends a patient just out of surgery in the Intensive Care ward of the hospital ship USS *Repose* (AH-16). *Photo courtesy of NARA.*

Bunk Bed Graffiti

As a newlywed Soldier heading to war in 1967, Zeb Armstrong was lovesick for his young wife, Billie. He feared he might never see her again. While lying in his bunk on board the ship making its way to Vietnam, Armstrong scribbled out his concerns on the bottom of the canvas bunk just inches above him. "Billie Armstrong, my dear wife," he wrote. He added "Black Cafe," a favorite hangout where he and Billie would go for burgers or hot dogs and dance to the jukebox. He included his name and hometown of Clover, South Carolina. And the last thing he wrote was, "Will I return?"

Armstrong was on board USNS *General Nelson M. Walker* (T-AP-125).* The journey took three weeks, and space was tight. The bunks—columns of tan canvases in stacks of four—were rigged to steel rods. A man lying on his back had only about 18 inches between his chin and the canvas bed above. Like Armstrong, hundreds of men used the canvas of the bunk above him to scribble out their anxiety, jokes, fears, and desires.

Air flights soon replaced *Walker* and her aging sister ships in transporting troops to South Vietnam. In December 1967, *Walker* was decommissioned and returned to reserve status. The bunk bed graffiti and everything else on the ship was left "as is" for about 30 years. When military historians Art and Lee Beltrone found it just before it went to the scrap yard, the ship was a virtual time capsule, capturing the stories of young men during wartime who left loved ones to serve their country. The historians wrote about it in *A Wartime Log and Vietnam Graffiti: Messages from a Forgotten Troopship*.

The war was the worst experience Armstrong ever had, says his granddaughter, Yuniqua Burris. "He felt he had to [serve] to protect his country, and he prayed a lot." His wife, Billie, remained his motivation, and wrote him often. After Armstrong returned to Clover, the couple raised seven children and founded a business. When Armstrong was diagnosed with cancer, Yuniqua asked him if he was afraid to die. He told her he wasn't. He had made it home from the Vietnam War, and had been able to live the life he had wanted with Billie, his dear wife.

— **Zeb Armstrong**
U.S. Army, Sergeant, 337th Signal Company (Tropospheric Radio Relay), 39th Signal Battalion, 1st Signal Brigade
Vietnam: 1967–68

*A reactivated WWII-era Maritime Reserve Fleet ship that transported troops to Vietnam from 1965–68.

I Considered Myself Pretty Lucky

Marines called the firebase camp at Con Thien, South Vietnam, "the Meat Grinder," remembers Joseph Prindle, "because we would average 800-900 rounds of [enemy fire] incoming every day—rockets, artillery, mortar rounds."

The day before Prindle and his buddy Jimmy Youngblood were scheduled for R&R, after six months in country, they were hit by shell fragments outside their bunker. "I got hit in the head, with shrapnel in my back and my left leg. Jimmy was killed, but I didn't know it at the time."

Prindle also didn't know the extent of his own injuries. He saw blood pumping against the wall, and not knowing it was his own, told the corpsman, "Doc, you'd better take care of the guy behind me, because I think he's bleeding to death." The main artery in Prindle's leg was cut, and the corpsman put on a tourniquet to try to stop the bleeding. The first medevac chopper was hit before it could land, killing everyone on board. A second chopper flew him to the hospital ship USS *Repose* (AH-16), off the coast of South Vietnam.

After a month, Prindle was transferred to Great Lakes Naval Hospital near Chicago.

At 135 pounds, he had lost so much weight his parents didn't recognize him. When they came to see him in the hospital, Prindle's mother walked right past his bed and kept calling his name, looking for him. "I said, 'Mom, it's me, it's OK.' And they cried, and I cried."

He spent 13 months at Great Lakes Naval Hospital near Chicago. Eventually, his leg had to be amputated.

Most of Prindle's fellow patients on the hospital's seventh floor had also been wounded at Con Thien. One night in 1968, as they watched the TV news, they learned that Con Thien was no longer considered a strategic position to be defended, and was abandoned in favor of a more flexible use of U.S. forces. The men felt betrayed and angry.

"The place became complete chaos," Prindle recalls. "People were throwing TVs and wheelchairs and anything not tied down through the windows of the seventh floor."

But most of the time, he says, there was a spirit of camaraderie among the recovering patients. They became proficient with their wheelchairs, racing them through the halls. "Ambulatory guys like me would go and feed the guys who were paralyzed. I learned to take blood pressures and temperatures, because the staff was short-handed and I was bored." Prindle thinks the long hospital stay helped him adjust to civilian life.

"A lot of Marines came right out of heavy combat, and within 36 hours they were on their front step," he says. "I had 13 months to decompress with people who had lost both arms, both legs, and in a lot worse shape than I was. So I considered myself pretty lucky."

— Joseph Prindle Jr.
U.S. Marine Corps, Corporal, radar operator, Headquarters, 12th Marines, 3rd Marine Division
Vietnam: 1967

The crew of U.S. Coast Guard cutter *Point Comfort* inspects a junk to make sure it contains no hidden contraband. *Photo courtesy of NARA.*

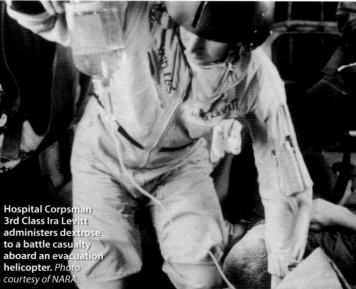

Hospital Corpsman 3rd Class Ira Levitt administers dextrose to a battle casualty aboard an evacuation helicopter. *Photo courtesy of NARA.*

The Wild Weasels

Rumor was, the attrition rate among The Wild Weasels—the guys who flew in the F-105 Thunderchief special electronic warfare planes designed to ferret out and destroy North Vietnamese surface-to-air missile (SAM) sites—was about 100 percent. As Lew Chesley discovered, that was because Weasels didn't have much of a chance of surviving.

Weasels were chosen from the best of the B-52 bomber and F-100F fighter pilots. When Chesley became a Weasel in 1966, he already had more than 2,000 hours flight time in B-52s as an electronic warfare officer. He was known as a "GIB" (guy in back), the guy who sat behind the pilot and used the plane's radar detection equipment and his expertise to locate the SAM sites down below.

After he arrived in country, Chesley and his fellow Weasels were filling out forms when a sergeant walked up, ripped off the back page of each form, and threw them in the trash. Chesley pulled out a crumpled paper and discovered it was what the pilots called "the dream sheet"—the form on which pilots indicated their preferences for their next assignments. When Chesley asked why he'd thrown away the papers, the sergeant explained: "We don't waste your time filling them out, and we don't waste our time keeping a file on them." It was expected, Chesley realized, that the Weasels would be shot down—and either die or become POWs.

Here was the warning the squadron commanders gave all new Weasels: "Make sure North Vietnamese civilians don't capture you, because they'll execute you. They will machete you to death, they will pitchfork you, hold you under water till you drown—whatever will kill you. You want to be escorted through the streets of Hanoi with a bayonet in your back, because that means the North Vietnamese are interested in keeping you alive. If you walk by a camera, face the camera, so the world will know that you are alive. Otherwise, you disappear."

Chesley was one of the lucky ones. He was never shot down, he was never a POW, and, after flying 100 missions, he came home alive.

— Lew Chesley
U.S. Air Force, Captain, electronic warfare officer, 354th Tactical Fighter Squadron, 355th Tactical Fighter Wing
Thailand: 1966–67

Three Air Force F-105 Thunderchief pilots enroute to bomb military targets in Vietnam pull up to a flying Air Force "gas station." The refueling aircraft is an Air Force KC-135 Stratotanker. *Photo courtesy of NARA.*

24367

10127

U.S. AIR FORCE

U.S. AIR FORCE

Two Piles of Marines

On the afternoon of April 25, 1967, after spending three months "out in the boonies," Bill Akers and his company were taken to Khe Sanh Combat Base (KSCB). They had no time to rest. His squad immediately boarded a helicopter and headed to help another company on Hill 861, northwest of KSCB. After landing, two platoons of Company K, including Akers' squad, began to climb the hill. They made it almost to the summit before settling in for the night.

"You could hear shooting at the top of the hill all night long," Akers recalls. "It wasn't that far away. The next morning before sunrise, we headed up the trail. Near the top we had to get down on our hands and knees and crawl because it was open, just some grass and scraggly trees. I had an automatic rifleman in front of me, but I was second in line, and just as we crawled out into the open, they started shooting at us."

Akers fired three rounds before his M-16 jammed. He dropped the rifle and started to throw grenades; the enemy retaliated by throwing some of their own. At one point, Akers saw four or five grenades bouncing around him and rolling down the hill. Eventually one exploded nearby, sending shrapnel into his leg, and then he got shot in his arm and chest.

"It was over for me that day," Akers recalls. "I just laid there; that's all I could do. I had a punctured lung, a sucking chest wound, and a large portion of my left arm was gone." The Marine behind him, still under the protection of the jungle, yelled at him to "get back here" so he could get help.

"There was no way I could crawl," he says. "Somehow that Marine got hold of me and pulled me back. Then he took my thumb and stuck it into the hole to plug up the hole in my chest and bandaged me up."

Two Marines loaded Akers on a poncho and tried to get him down the mountain. "They ran about 10 yards, then the shooting would start, and they'd have to get down and drag me," Akers remembers. That happened three times before the two Marines were shot—plus Akers got shot in his right leg.

Ultimately, Akers said, another Marine grabbed his poncho and took off. "That guy couldn't have weighed more than 145 pounds, and I weighed about 200," Akers says. "But he flew down that mountain."

Akers remembers the landing zone and seeing two piles—one pile of dead Marines and one pile of wounded. He still can't talk about it without getting choked up.

Akers spent eight months at Walter Reed Army Medical Center in Bethesda, Maryland. He recovered from his physical wounds, but he's still working on the emotional ones. "That's the reason I go to PTSD counseling," he explains. "I'm still trying to reconcile it all."

— **Bill Akers**
U.S. Marine Corps, Company K, 3rd Battalion,
3rd Marines, 3rd Marine Division
Vietnam: 1967

The Call of the Wild

Vietnam's extremes of nature held both beautiful and terrifying surprises for William Mershawn. He spent most of his time out in the field, patrolling under triple-canopy jungles, through 15-foot-high elephant grass, and across mountainous vegetation.

"The most memorable sights were the flights of butterflies," he says. "A cloud of a mile or two would just appear and fill the air with yellow and black and white wings. When you could stop and get away from the war for a minute, it was such a beautiful area with green trees and flowers and hills." The natural beauty stood out in sharp contrast to the death and destruction of combat.

Some of the not-so-pleasant wildlife included large scorpions and 18-inch centipedes in the A Shau Valley. One man in the platoon tripped over a python.

"Fortunately, it had just swallowed a small animal and didn't wrap him up," Mershawn says.

One night, during Mershawn's second tour of duty, the platoon had just set up an ambush position for a surprise attack when they heard sounds of movement approaching. Suddenly, a full-grown tiger came out of the jungle. "I guess he could smell us," Mershawn says. "He looked at us as he kept nosing closer and closer." No one dared move.

"I had already put my weapon on full automatic, and we were out hunting a large North Vietnamese unit," he explains. "If you shoot, you give yourself away. So we were very hesitant to shoot." The tiger inched closer and was only about six feet away from the platoon. Thinking quickly, the Soldier nearest Mershawn picked up an aerosol can of bug spray and sprayed the tiger in the face. The tiger shook its head and looked back at the Soldier, sniffing. The Soldier sprayed the tiger again, and the animal finally turned and left—much to the platoon's relief.

Being out in the wild brought out the best in his men, Mershawn notes. "When you're in combat, after about the first 48 hours, the facade that most people have is stripped away. The naked personality really comes out and they are so likable. They were some of the finest Americans I've ever had the privilege to know."

— **William Ray Mershawn**
   U.S. Army, Sergeant, infantry, 2nd Battalion,
   12th Cavalry, 1st Cavalry Division (Airmobile)
   Vietnam: 1965–66, 68–69

Sergeant Curtis E. Hester, assistant patrol leader, Company D, 151st Infantry, Indiana ARNG, positions to fire his M-16 rifle against the enemy. *Photo courtesy of NARA.*

Well, It *Was* Paper

When Top Secret papers go missing, the Army will employ heroic measures to get them back, as Michael Gene McCracken discovered one day in South Vietnam.

McCracken's job was to repair and maintain the equipment used to transmit and store encrypted radio messages, to keep it secure from the enemy. While stationed at the 330th Radio Research Unit at Pleiku, the entire MACV communications network underwent a simultaneous, one-time code change. The six-month supply of codes for all U.S. military units worldwide were printed up and flown in to each unit on the same day, so the change could take place at the same time in every unit.

"They had a group of guys who were going to the airport to pick up all these boxes of codes that had all the settings for the cryptography gear," McCracken recalls. The Soldiers put the boxes in the back of a three-quarter-ton truck. But they neglected to shut the tailgate.

"When they got back to the unit, it was dark, and they parked in the security area, and didn't think to check if the boxes were still there," says McCracken. "The next morning someone came by and wanted to get the codes and noticed they weren't in the truck. Some time went by, and suddenly they realized they had lost all the Top Secret codes. That was a severe violation, because the codes were compromised."

They began desperately searching for the boxes, retracing the truck's path to see where they might have fallen out. The next day, he recalls, "in walks an ARVN [Army of the Republic of Vietnam] lieutenant, who can speak and read some English. He hands [the supervising officer] a piece of paper marked 'Top Secret,' and asks if it belongs to the unit. [The officer] said yes and asked where the rest of the papers were. The lieutenant said, 'My unit found it, and we're using it for toilet paper.'"

The NCO and a few of the men "got the unenviable job of retrieving all the used code papers, and accounting for every one of them," McCracken recalls. "They were using tongs and wearing gas masks. It was a very serious thing, but it was also rather funny."

— **Michael Gene McCracken**
U.S. Army, cryptographic repairman, 330th Radio Research Unit, 196th Infantry Brigade (Light), 23rd Infantry Division
Vietnam: 1967–69

His Death Was Greatly Exaggerated

Although his name is etched on the Vietnam Memorial Wall as a Soldier killed in combat, David F. Kies is alive and living in Wisconsin.

In January 1967, he was on a reconnaissance team of the 173rd Airborne Brigade, in thick jungle near the Vietnam-Cambodian border. His team was patrolling for booby traps when Eric Zoller, just in front of him, triggered an enemy mine. Kies heard a sudden boom.

"I was thrown up in the air, and when I tried to get up, I couldn't," Kies recalls. "I looked down and one of my legs was gone. And the other one was broken."

A medic used Kies's belt as a tourniquet. There were no medevac helicopters in the area, so a colonel on his team brought in his own helicopter to fly him to a MASH* hospital. His remaining leg was amputated. He was eventually sent to Walter Reed Army Medical Center, Washington, D.C., where he raced wheelchairs with other amputees. Later, he went to the Veterans Administration hospitals in Madison, Wisconsin, where he was fitted for artificial limbs.

Kies finished college, went to work, got married, and had five children. Meanwhile, due to a paperwork error, or perhaps because he wasn't brought in by medevac helicopter, he was listed as KIA. In the mid-1980s, someone saw his name on the Wall and alerted him. "Well, we're dealing with the government," Kies explains, unbothered. "I think there were 34 mistakes out of 58,000 names, so not too bad. Probably not bad at all."

Since his comrades thought he was dead, more than 30 years passed before he was invited to reunions. In 2000, a guy from Kies's unit came across a name and phone number on the Internet for a David Kies, and decided to call. He reached Kies's son, who has the same name.

"He told my son, 'I knew your dad, and I want to talk to you about him,'" Kies explains. "My son said, 'Why don't you just call him?'" The war buddy, surprised, asked, "What do you mean, call your dad? Isn't he dead?!"

Kies was soon flooded with calls from former comrades. And for the first time in 33 years, he attended the 173rd Airborne reunion, where men were surprised he wasn't in a wheelchair. "So I went from being dead, to in a wheelchair, to actually walking," he says, pleased.**

— **David F. Kies**
U.S. Army, Sergeant, 173rd Airborne Brigade
Vietnam: 1966–67

*Mobile Army Surgical Hospital, a U.S. Army medical unit serving as a fully functional hospital in a combat area of operations.

**Excerpts of this story are credited to a James Kurtz interview in 2002 for the Wisconsin Veterans Museum Oral History Program in Madison, Wisconsin.

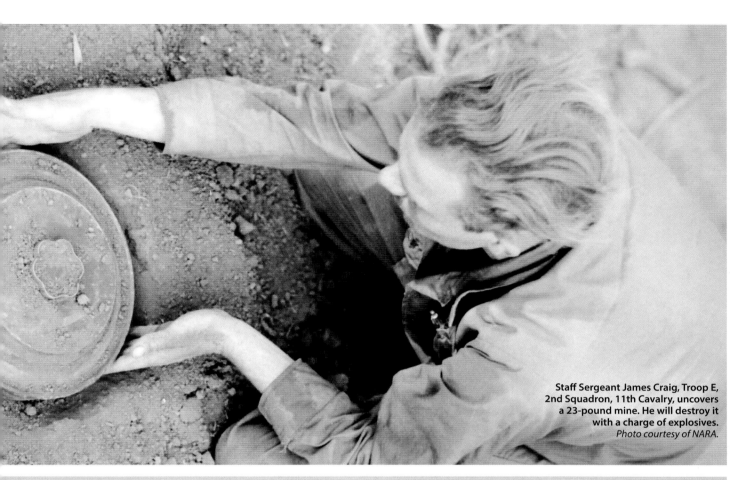

Staff Sergeant James Craig, Troop E, 2nd Squadron, 11th Cavalry, uncovers a 23-pound mine. He will destroy it with a charge of explosives.
Photo courtesy of NARA.

Yankee Air Pirates

In July 1967, a photo reconnaissance mission over North Vietnam returned to the USS *Constellation* (CVA-64) with images of a surface-to-air missile (SAM) site about 10 miles south of Hanoi—in the middle of a soccer stadium.

It was a perfect target, says Robert Dunn, commanding officer of Strike Fighter Squadron 146, the "Blue Diamonds," stationed on board the *Constellation*. The Blue Diamonds "salivated at the chance" to destroy the missiles that had caused so much damage to U.S. carrier squadrons. A dozen A-4C Skyhawks, four A-6E Intruders, and four F-4J Phantoms soon took off from the ship and headed for the enemy coastline.

"The flak grew more intense as we weaved and jinked our way in," Dunn remembers. "Screeching electronic cockpit warning systems alerted us that we were being targeted by radar-guided anti-aircraft guns and that surface-to-air missiles were headed our way."

Amid shelling from 57mm and 105mm guns, Dunn's Blue Diamonds spotted the target. "The missile battery sat as pretty as you please, right in the middle of the vacant sports stadium," he says, and the U.S. planes went for it.

A photo-reconnaissance plane following close behind the strike group brought back proof that the bombs had destroyed the North Vietnamese cache of SAMs. Several days later, though, Dunn learned about the lead story in a Hanoi newspaper: "Yankee Air Pirates Destroy Schoolchildren's Football Stadium." There was no mention of the missiles.

According to historians of the war, the North Vietnamese preferred placing their SAM batteries close to nonmilitary sites, hoping that would keep U.S. planes from risking collateral damage. If that didn't work, at least Hanoi would have good propaganda photos. The newspaper headline, Dunn says, "was just one more episode in North Vietnam's effort to influence anti-war sentiment in the United States by painting the air campaign as a criminal enterprise."

— **Robert F. Dunn**
U.S. Navy, Commander, pilot, Attack Squadron 146, USS *Constellation* (CVA-64)
South China Sea: 1967

Dunn retired from the U.S. Navy as Vice Admiral in 1989.

Putting Spirits to Rest

In 2004, on a trip to Vietnam, Wayne Karlin hiked up Marble Mountain. In his backpack were several sticks of incense, and when he reached the very top he lit them. Then he said a prayer for Jim Childers.

It had taken Karlin several trips to Vietnam to work up the courage to visit Marble Mountain. During the war, he had served as a helicopter gunner stationed at the air base there. In 1967, in his final few days in country, he had been assigned to a routine supply mission, but at the last minute his friend, Jim Childers, had volunteered to take Karlin's place. Childers was getting married soon and wanted a twentieth mission so he could earn his combat aircrew wings and wear them at his wedding. It was a fatal flight. Childers was hit by a bullet as the helicopter flew over enemy positions, and he died on the operating table.

So the incense 27 years later was for Childers. But also, Karlin says, "for all the people who died in our places and in our names."

Now a professor of Languages and Literature at the College of Southern Maryland, Karlin has written novels, short stories, and screenplays about Vietnam and the war, and has edited anthologies of Vietnamese writers. In 2005, having heard of Karlin's work, an Army veteran sent the professor a request. Could Karlin help him locate a family in Vietnam? The veteran's name was Homer Steedly Jr.

Later, Karlin chronicled this story in his book *Wandering Souls*. A short version of it goes like this: In 1969, when 1st Lt. Homer Steedly was out on patrol, he came upon a North Vietnamese soldier walking alone on the Ho Chi Minh trail. Both men were startled and both reached for their guns, but Steedly shot first, killing young medic Hoang Ngoc Dam. Searching the dead soldier's uniform, Steedly found documents and a notebook full of poems, medical drawings, and games of tic-tac-toe; he thought they were interesting, so he mailed them home to South Carolina. Thirty-six years later, he discovered them in his parents' attic.

Meanwhile, Dam's family had been trying for years to locate Dam's grave. They worried that because his body had never been sent back to his home, he was what the Vietnamese call a "wandering soul," never at peace. So when Karlin tracked the family down, they were eager to get Dam's artifacts back, and thus finally put his soul to rest. They even wanted to meet Steedly, but he sent Karlin to Vietnam in his place. Steedly also sent a letter that included this wish: "Maybe some day humanity will have the wisdom to settle conflicts without sending its youth to kill strangers."

The notebook was placed on the family altar, and incense lit to welcome Dam's soul home.

Three years later, Steedly was finally ready to meet the family, and Karlin flew back to Vietnam with him. Again they lit incense, and then traveled with the family to Pleiku to find Dam's remains and bring them home for a funeral. Steedly was one of the pallbearers.

An estimated 300,000 Vietnamese soldiers who served in the war are still missing, all of them wandering souls. But so many American veterans—still living—are wandering souls, too, Karlin says; Soldiers like Steedly, who returned physically, but could not truly "come home" until he had mentally dug up the past.

"Recovery from trauma," Karlin says, "involves disinterring all the bad stuff you've buried in yourself. You have to bring it out into the light, you have to mourn what you have lost, and you have to find a new way of living with it. And then you can put it to rest."

— **Wayne Karlin**
 U.S. Marine Corps, Sergeant, 1st Marine Air Wing
 Vietnam: 1966–67

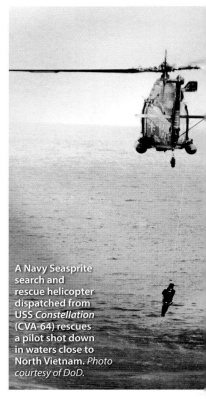

A Navy Seasprite search and rescue helicopter dispatched from USS *Constellation* (CVA-64) rescues a pilot shot down in waters close to North Vietnam. *Photo courtesy of DoD.*

A 1st Infantry Division "tunnel rat" wearing a gas mask is helped out of a Viet Cong tunnel after completing a search. *Photo courtesy of NARA.*

A member of a U.S. Navy SEAL team uses caution as he watches for any movement in the thick wooded area along a stream. He carries a Stoner 63A light machine gun.
Photo courtesy of NARA.

It Means Everything to Us Vets

After fighting in Vietnam for a year, John Garcia came home, got married, and enrolled at the University of Albuquerque. The first day of class, one of his instructors announced, "If there are any Vietnam vets in here, I want you out of here." Garcia sat there, astonished. "Then the guy next to me leans over and asks, 'Hey, John, are you a vet?' When I told him I had just got back, he replied, 'Don't say anything right now, but meet us tonight at 8.'"

Garcia met his new friend that night, along with a dozen other guys—all Vietnam vets. "They told me not to tell anyone I was a vet," he remembers. "And so I learned very quickly—I was 19—don't say nothing. Just keep quiet. I've always said, Vietnam I could handle—it's my coming home that was the toughest thing for me."

Garcia went to the local VA office to file for his benefits, but the place was full of angry vets, so he left. It took him 30 years to return. "I finally went back because my wife told me she was tired of chasing these [PTSD] dragons with me," he explains. "'If you don't go to the VA for yourself, go for me,' she said."

So he started his healing journey. He became deeply involved in helping and serving fellow veterans from all wars, but especially those from Vietnam. When the Vietnam Memorial was dedicated in 1982, Garcia raised enough money to take 135 fellow vets back to Washington, D.C., for the dedication.

"I remember we were walking down the main street with people on the side saying, 'Welcome home, thank you,'" Garcia remembers. "That was the first time any of us had ever heard that. 250,000 Vietnam War vets showed up for the dedication of the Wall, and we were a ragtag group of guys. But we were together again, trying to find each other and find names on the Wall." It took Garcia five tries before he finally got up enough nerve to see the Wall up close. "When I did, it swallowed me," he remembers. "I just bawled like a baby."

Two years later, in 1984, a statue group of three Soldiers and a flagpole were added to the memorial area. The American flag is flown there 24 hours a day. The flag—and what it represents—is sacred to Garcia and tens of thousands of his fellow Vietnam War vets. Before the Wall was built, Garcia worked to build a small chapel in Angelfire, New Mexico, with a man whose son had died in Vietnam. The man wanted to build it to honor his son. "It became kind of a monument for vets," Garcia says.

One day he got a call from a Vietnam War vet, Mike McWaters, who had a flag he wanted to hang at the chapel, now called the Vietnam Veterans Peace and Brotherhood Chapel. "When we met, he reached into his jacket and pulled out this dirty, torn American flag," Garcia recalls. "And then he told me a story."

McWaters was a Marine at the Khe Sahn Combat Base, and during the famous battle there in 1967, a mortar hit a flagpole on the base and the American flag fluttered to the ground. "McWaters ran through enemy fire to retrieve that flag in the mud," Garcia says. "I asked him why he would risk his life to save a flag, and he said, 'Everything I believe in— my family, my church, my school, my honor—everything was in that flag there in that mud, and I had to save that flag.'" McWaters' flag now hangs with honor in the chapel.

"So when you ask what the flag means—it means everything to us vets," Garcia explains.

— John Garcia
U.S. Army, Sergeant, infantry, 1st Battalion, 14th Infantry; Headquarters Company, 4th Infantry Division
Vietnam: 1969–70

20th-century angel of mercy: D. R. Howe treats the wounds of Private First Class D. A. Crum, Company H, 2nd Battalion, 5th Marines during Operation Hue City.
Photo courtesy of NARA.

Getting Shot At Never Gets Routine

During Thomas Schottenbauer's second tour of duty in Vietnam, his ship—the guided missile cruiser USS *Canberra* (CA-70)—was on Condition 3 watch. "That meant one-third of the crew was always on watch," he explains. "You were on for four [hours], off for eight, on for four, off for eight—and that would continue all while you were there." Schottenbauer served his watch in weapons control central, the ship's nerve center. "All the weapons systems get their command to fire from that particular room," he says.

During what they called "Moses cruises" (40 days, 40 nights), things got pretty intense. "When you live that close together, and that tight, you get to be good buddies with everyone," Schottenbauer recalls. "But on the Moses cruises, tempers would start getting short." To let off some steam, the crew had "smokers," or boxing matches. Schottenbauer participated in two, and won both in knockouts.

Schottenbauer saw plenty of action on his second tour, which lasted eight months. "We were working a lot north of the DMZ, taking out missile sites and gun placements in North Vietnam," he says. "We got shot at and hit quite a bit. We actually broke records set in World War II, of ammunition expended in the shortest period of time."

Schottenbauer had a close call on March 2, 1967, when two enemy shells hit the ship directly in the area where he was working. He never got used to being shot at. "Something happens to you when somebody shoots to kill you," he concludes. "You never think the same again—your whole attitude for the rest of your life has changed."

— Thomas Schottenbauer
U.S. Navy, Petty Officer 3rd Class, gunner's mate, USS *Canberra* (CA-70)
Vietnam: 1965–66, 66–67

The Long Road to Citizenship

Victor Macias was an American hero long before he became an American citizen. After moving with his family from Mexico to the United States when he was 15, he had legal alien status, which meant he could be drafted into the Army. In training, his drill instructor, a Sgt. Vasquez, took him aside and said, "Macias, if you want to be noticed by the higher-ups, you have to be twice as good as the white boys."

In South Vietnam, he became a helicopter scout observer, riding in the left seat next to the pilot in a Hughes OH6A "Loach"* helicopter. The scout ship would fly at treetop or grass level, as Macias looked for signs of the enemy—and took enemy fire.

On May 18, 1968, his patrol got a call that another 17th Air Cav scout helicopter had been shot down. "We flew to the area and found the ship burning, tipped on its right side where the pilot was," Macias remembers. The scout observer, Henry Jackson, had crawled out of the helicopter. Macias's pilot landed the Loach between two bomb craters, and Macias ran down the hill to Jackson, with bullets whizzing all around him. "I got to Henry and put him on my back, but the earth was so loose and we were receiving fire, so I set him behind a rock to give him cover," he explains. "Jackson was badly burned, and when I put him down, I realized both his legs were broken."

When the medevac helicopter came, Macias helped load Jackson, then headed back up the hill, still under fire. In the meantime, the Loach had been damaged by enemy fire— Macias found his own pilot lying in a bomb crater, shot in the foot. Macias helped the pilot get to the bottom of the hill, where the chopper returned for them.

Macias hoped his Vietnam War service would help him gain U.S. citizenship. When he came home, he went to the Immigration and Naturalization Service with his Army documents and medals, including a Purple Heart, Silver Star, and Air Medal for Heroism. An official told him, "Buddy, with your medals and 50 cents, you could buy yourself a cup of coffee. You are no better than those people in line from other countries." Macias was so angry, he didn't apply for citizenship again until 1985—and finally became a citizen on Sept. 2, 1986.

In 2003, Macias attended a 7th Squadron, 17th Air Cavalry reunion, where the five scout survivors shared memories and drank to those who never returned. "They referred to me as a brave man," he says. "But all I did was fight for my friends who were counting on me, and help my brothers in need."

— **Victor Macias**
   U.S. Army, Sergeant, scout observer,
   Troop A, 7th Squadron, 17th Air Cavalry
   Vietnam: 1967–68

*Light Observation Aircraft

The Red-Headed Stranger

On November 16, 1967, Jim Deister's demolition team landed in a hail of bullets in the Plain of Reeds, a wetland depression stretching west of Saigon towards Cambodia. The next day, the team blew up a downed AH-1 Cobra gunship, then moved by helicopter to a remote fire-support base. They set up with the infantry reconnaissance platoon by a large canal, filling and stacking sandbags for protection.

When Jim heard the whoosh of a mortar shell during the night, he knew what was coming. A storm of mortars rained down on the American troops, while hundreds of North Vietnamese soldiers led the charge, carrying AK-47 assault rifles. "You die tonight, GI!" they kept screaming.

PFC Deister, who didn't know the infantry or artillery guys, was desperately trying to find his demolition buddies. A flare lit up the sky, and in the distance he saw his teammate Jim Dailey lying face down, with blood smeared on the back of his neck. Deister was making his way towards his friend when a fragment hit him in the chest—and then he took a bullet in the head.

"Things got hazy about that time," he recalls, "and what I remember was like watching a slide show, where you see a slide, then miss two or three, then see another. But nothing makes sense." He remembers one face: a redheaded Soldier peering through the bushes. And he remembers floating on his back in the canal, watching flares light up the sky.

He woke up in a hospital in Japan—he'd been brought in with a six-inch wound on the right side of his head and brain matter protruding from the hole. Over the years, Deister kept wondering how he'd survived that night in 1967. He finally got an answer in 1988, when by chance he picked up a book about Medal of Honor recipients. A familiar date and location grabbed his attention: it was an account of the offensive that had almost killed him. He then read how Sammy Lee Davis, an artillery Soldier who later received the Medal of Honor, had crossed the canal to assist three stranded U.S. Soldiers— one of whom had been shot in the head.

Deister contacted Davis, who turned out to be the red-haired fellow he remembered. He learned that one of the Recon Platoon Soldiers had found him struggling on the trail and dragged him to the canal bank. The infantry medic had put him "in the pile with the dead guys" until someone saw him move. Thanks to strangers—his fellow Soldiers—Deister came back to his wife and baby daughter Jamie, born just 19 days before he was deployed to Vietnam.

— **Jim Deister**
   U.S. Army, Private First Class, explosive ordnance disposal,
   3rd Platoon, Company C, 15th Engineer Battalion
   Vietnam: 1967–68

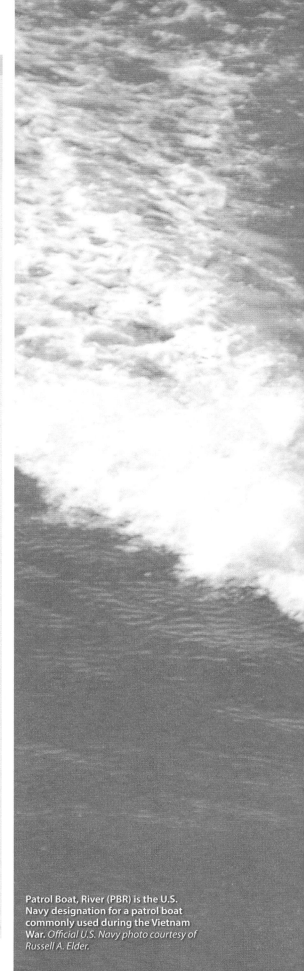

Just About Anything to Stay Alive

Russell Elder was moving downriver on a U.S. Navy river patrol boat near Can Tho, South Vietnam, when an enemy rocket blasted into the boat. "Luckily, we had a boat captain who knew what the hell he was doing," recalls Elder, who was below deck when the rocket exploded. "He hit the throttle at just the right moment, which kicked the back of the boat down—and we caught the rocket above the waterline instead of below it, so the boat didn't take on nearly as much water."

Elder, a Navy photographer whose assignment was to capture as much of the war as possible, ran up top and started shooting pictures. "When you take photos, you feel a little separation," he explains. "You're watching things happen, and I didn't always feel like I was part of the action."

But this day was different. Adrenaline flowing, Elder was more aware than ever of the danger he and his fellow Sailors were in. The crew headed to a nearby riverbank, grounded the damaged boat, and waited for another boat to pick them up.

"I was sitting there in the front of the boat, holding a .45 [pistol] in my hand and watching as we hit the shore," he says. "My brother had come back from Vietnam a few years earlier, and he told me about guys who had to shoot women and children who were wired with explosives. I couldn't understand that at the time, but as I sat there in that boat heading toward shore, I realized that if someone had crawled out of the bushes in diapers, I would have shot. I suddenly found out I would do just about anything I needed to do to stay alive."

It wasn't until after the crew had returned to base and a Navy corpsman offered to buy Elder a beer to celebrate his first Purple Heart that Elder realized he'd been burned in the attack. "I don't know how or when it happened," he says. "But they bandaged it up, and I stayed. My commander called and asked if I wanted to get out of there, but I told him, 'Hell, no.'"

Ultimately, Elder did go home injured. On assignment aboard a Navy "swift boat,"* Elder remembers the sun coming up on an "absolutely gorgeous Vietnam morning, when we heard a swarm of bees—only they weren't bees. They were bullets, and I got hit. The bullet damaged my arm, and I said the two dumbest things I've ever heard anyone say: 'I'm hit.' And then, 'Why me?'"

Elder had earned his second Purple Heart and was sent home. "You don't really understand the pressure you're under until it's gone," he reflects. "When they put me on that C-141 [transport aircraft] and flew me to Japan, a corpsman wheeled me out on a gurney, and he said, 'It's okay. You're safe now.' And the dam broke. It still does every time I talk about it."

— Russell A. Elder
U.S. Navy, Photographer's Mate 1st Class, photojournalist,
Pacific Fleet Combat Camera Group; I Corps
Vietnam: 1964–65, 1968–69

*"Swift boat" was the unofficial name given to the U.S. Navy's Patrol Craft Fast (PCF)—all-aluminum, 50-foot-long, shallow-draft vessels used to patrol South Vietnam's coastal areas and interior waterways. A second type of small boat, Patrol Boat, River (PBR), had a fiberglass hull.

Patrol Boat, River (PBR) is the U.S. Navy designation for a patrol boat commonly used during the Vietnam War. *Official U.S. Navy photo courtesy of Russell A. Elder.*

U.S. Huey (UH-1) helicopters bring in a second wave of troops to a "green" LZ. *Photo courtesy of NARA.*

1968
Year of the Offensive

The year began with a traditional holiday "truce"—but that ended on January 31, when the North Vietnamese Army and Viet Cong launched coordinated attacks on more than 100 cities and towns across South Vietnam.

Named for the Vietnamese lunar new year, the Tet Offensive took many by surprise, but the Communist effort to spark a "popular uprising" against the South Vietnamese government failed. After several weeks of heavy casualties on both sides, South Vietnam reclaimed its captured territory.

The Tet Offensive was a turning point. Even though it was a military victory for the United States, it turned out to be a major North Vietnamese propaganda victory, as America's first "TV war" brought unsettling images and increasing doubt into American homes and college campuses. Even as the U.S. won other hard-fought victories, the American media were full of talk about *quagmire, credibility gap*, and *de-escalation*. In May, U.S. and North Vietnamese representatives met secretly in Paris without result. Soon after, General Creighton Abrams became the new Military Assistance Command, Vietnam (MACV) commander. Under heavy pressure from the anti-war movement, and losing Congressional support, President Lyndon Johnson announced he would not seek a second term. Later that year, in an unsuccessful effort to restart the stalled peace talks, Johnson ended the bombing of North Vietnam.

Soldiers from the 9th Infantry Division move upstream as they patrol a river in South Vietnam.
Photo courtesy of NARA.

TIMELINE 1968

January 30, 1968
Viet Cong launch "Tet Offensive" on South Vietnam; 268,800 Communist troops attack 105 Vietnamese cities and towns, leaving 81,000 dead and 350,000 homeless.

April 4, 1968
Dr. Martin Luther King Jr. is assassinated.

May 10, 1968
Peace talks start in Paris.

June 1, 1968
Simon & Garfunkel song *Mrs. Robinson* hits No. 1 on music charts.

June 6, 1968
U.S. Senator Robert F. Kennedy is assassinated while campaigning for president.

June 27, 1968
Marines begin withdrawing from Khe Sanh.

July 1968
General Abrams takes control of Military Assistance Command, Vietnam (MACV) from General Westmoreland.

August 26–29, 1968
Anti-war riots in Chicago at DNC Convention.

October 16, 1968
African-American track stars protest racism in the U.S. by giving "Black Power" salute during national anthem at the 1968 Summer Olympics in Mexico City.

October 31, 1968
President Johnson orders cessation of bombing of North Vietnam. North and South Vietnam take part in the Paris Peace talks.

November 5, 1968
Richard M. Nixon is elected president.

December 1968
536,000 U.S. military personnel in Vietnam; casualties reach 30,000.

A "leatherneck" dashes forward with his M-79 grenade launcher as the Special Landing Force battles a North Vietnamese battalion. *Photo courtesy of NARA.*

He's My Brother

In 1968, Chuck Hagel and his younger brother Tom became the only known American siblings to serve in the same infantry squad during the Vietnam War. The two brothers fought side by side in the steaming jungles of the Mekong Delta. They walked point together, watched comrades die in battle, and between the two of them, sent five Purple Hearts home to their mother. They also saved each other's lives.

Tom, two years younger than Chuck, saved his older brother first. Normally the Hagel brothers walked point, but one morning in 1968 they had rotated to the rear as their column of Soldiers crept through the jungle. The Soldier who took their place that day met instant death as he stepped on a huge land mine. Flying shrapnel from the mine ripped through the squad. It hit Tom's arm, but a bigger chunk lodged in Chuck's chest. Ignoring his own wound, Tom frantically wrapped compression bandages around Chuck's chest to stop the blood, praying his older brother would live long enough to make it out of the jungle alive. He did. And then returned to duty after he healed.

Only a few months later, Chuck was riding on top of the last M113 armored personnel carrier leaving a fierce firefight. A hidden Viet Cong waited for the lone APC and blew it up with a command-detonated mine. Instantly aflame, his left side injured and his face badly burned, Chuck ignored his wounds and frantically tugged Tom, the turret gunner, out of the blasted and burning vehicle. "With all the ammunition we had, it would just blow," remembers Chuck. "I grabbed Tom, and he was dead weight, blood pouring out of his ears. I thought for sure he was dead."

Chuck shoved Tom off the APC, then fell on top of him to protect him as the enemy opened fire. Flames from the burning vehicle shielded the brothers until the trucks up ahead turned back to save them.

Twenty-nine years later, in 1997, the two brothers stood together in the U.S. Capitol as Chuck was sworn in as the new senator from Nebraska.

— **Charles "Chuck" Hagel**
   U.S. Army, Sergeant, squad leader, 2nd Battalion, 47th Infantry, 9th Infantry Division
   Vietnam: 1967–68

— **Thomas "Tom" Hagel**
   U.S. Army, Sergeant, 2nd Battalion, 47th Infantry, 9th Infantry Division
   Vietnam: 1968–69

Chuck Hagel served two terms as senator and then was appointed Secretary of Defense (2013–14) under President Barack Obama. Tom Hagel is an emeritus professor of law at the University of Dayton School of Law.

(Excerpts) Myra MacPherson, "The Private War of Chuck and Tom Hagel," *Salon*, April 30, 2007, http://www.salon.com/2007/04/30/hagel_brothers/. MacPherson is a former *Washington Post* reporter and author of five books, including the Vietnam War classic *Long Time Passing: Vietnam and the Haunted Generation*, Indiana Univ. Press, 1984.

Members of a Long Range Patrol Team open fire against the enemy. *Photo courtesy of NARA.*

Members of a U.S. Navy SEAL team man their weapons as they prepare to come ashore from a river patrol boat (PBR Mark 1) at an operation site. *Photo courtesy of NARA.*

Loss of Reality

Reconnaissance scout Philip Randazzo had been in the jungle for four long months with his unit of two tanks and a few track* vehicles. The men conducted search-and-destroy operations, set up ambushes and listening posts, and crawled into tunnels to locate and destroy the elusive enemy.

On January 30, 1968—the day before Tet, the Vietnamese New Year—the unit was pulled back to base camp in Cu Chi. Unexpectedly, the camp was heavily mortared as the enemy launched a ruthless offensive. With no rest, Randazzo's platoon was off again, joining other 3rd Squadron vehicles on the highway to reinforce security at Tan Son Nhut Air Base near Saigon. As the column approached the airport, a relieved Randazzo took off his steel pot, ammunition bandoleers, and flak jacket.

Suddenly, NVA soldiers hiding in the village attacked. An RPG blew Randazzo off his track and onto the highway. With no weapons or ammunition, he crawled into a ditch. Within a couple of minutes, everyone on the lead tank and the first APC was dead or near dead, including his captain and the forward observer. As four NVA soldiers advanced across the road to clear the ditch, Randazzo prepared to die—but just then a U.S. Soldier shot at the enemy with his M60 machine gun. Randazzo and another survivor ran to one of the disabled tracks and grabbed an M-16 rifle, an M-79 grenade launcher, and ammo.

For the next 10 hours, Randazzo stayed in that ditch, surrounded by wounded and dead Soldiers and trying to keep the NVA at bay—all the while praying for artillery support that never came. A couple more U.S. Soldiers made it to the ditch, carrying additional weapons and ammo. Then a civilian Vietnamese man dressed in a dark suit and white shirt tried to cross the road; he was shot by the NVA and fell two feet from the ditch. His distraught wife ran toward him with one child in her arms and two others by her side; they too were hit by enemy fire.

A track finally managed to move toward the airport, breaking the enemy roadblock. Despite shrapnel wounds in his head, Randazzo loaded the dead and injured onto the track, then ran back to the ditch.

That was only the beginning of the Tet Offensive, which lasted almost 30 days. Randazzo says the rest of the week around the air base "was pretty rough," as they removed the bodies and took care of the wounded. Images from that first ambush will never be erased. "Combat veterans have to live with those kinds of memories," he says. "There was a feeling of loss of reality, like we no longer existed."

— **Philip Randazzo**
U.S. Army, Specialist 4, reconnaissance scout, 3rd Squadron, 4th Cavalry, 25th Infantry Division
Vietnam: 1967–68

Armored vehicles that use tracks rather than wheels to travel. In the Vietnam War, "track" most commonly referred to M113 Armored Personnel Carrier (APC) and M577 Command Vehicle.

We Wouldn't Leave Him Behind

One humid dawn in 1968, Army pilots from the 188th Assault Helicopter Company flew four UH-1 Hueys on what seemed like a routine mission into Cambodia. The pilots flew low in a trail formation to spread out the telltale noise of the Hueys and confuse armed enemy soldiers in the jungle below. One of the pilots was 20-year-old Chief Warrant Officer Kjell Tollefsen, who enlisted straight out of high school. "With the war escalating, they allowed high school graduates who could pass psychological and physical tests to take a shot at it," Tollefsen explains.

It was a special operations insertion, so only two of the Hueys that morning, including Tollefsen's, carried troops; the other two flew empty as backup. The pilots landed the troops, then flew back to camp to wait a day for the retrieval call. "It was only a couple of hours when they called for immediate extraction; they were under heavy fire," Tollefsen recalls. "We had inserted them in a battalion-sized North Vietnamese unit. There were already two [American] KIAs and three or four wounded."

Tollefsen and his crew were half a kilometer away from the retrieval site when his Huey was first hit by enemy fire; they kept going. They hovered as troops loaded the KIAs and wounded into his Huey. "On takeoff, we were taking hits the whole time, warning lights were going off, and we were losing oil pressure," Tollefsen says. "We were losing power."

An officer in the command and control ship orbiting above told Tollefsen he could either return to a tiny jungle opening or cross the border and attempt to reach a larger clearing a mile farther into Cambodia. Tollefsen chose the clearing. "When I was just over it, the engine stopped," he recalls. A controlled crash left the Huey on the right side of its nose, with the occupants scattering into the tall grass for cover. Regrouping, they realized the door gunner was missing.

"We went back and saw him flailing, [his] head pinned under the top part of the fuselage," Tollefsen says. The men, now vulnerable to enemy fire, worked to free their friend. "Our wingman landed next to us, and the special forces guys jumped out of the helicopter and set up a perimeter to try and hold off the bad guys. We tried to lift the fuselage off him for 30 minutes. Control said to leave him behind. We wouldn't do it."

Quick thinking prompted Tollefsen to cut the gunner's helmet strap—one hard pull freed the man, leaving behind a chunk of his ear. "He broke a collarbone, and they sewed his ear back together, what was left," Tollefsen says. "Later he came back to the unit. They say any landing you can walk away from is a good one. That was as close as you want to get to not coming home."

— Kjell Tollefsen
U.S. Army, Chief Warrant Officer, pilot,
188th Assault Helicopter Company, 269th Aviation Battalion (Combat)
Vietnam: 1966–68

Members of Company C, 2nd Battalion, 503rd Infantry (Airborne),173rd Airborne Brigade charge from their Hueys toward a tree line in a heliborne assault. *Photo courtesy of NARA.*

Jerry, You Need to Do Something

The sound of helicopters is permanently burned into Jerry Murray's mind. "You will never forget the whop-whop-whop-whop of the helicopters," he says. Even years later when hearing a civilian hospital's Life Flight helicopter, Murray's mind flashes back to Vietnam.

It was March 1968, and snipers had pinned down Murray's infantry company at a small South Vietnamese village known as Lam Son East. Lieutenant Sisk, a platoon leader, had been wounded and was lying in a trench. "The radioman was about five or ten feet away trying to get a medevac helicopter and let the rest of the company know," he recalls. Murray was about 20-25 yards away, and could see that Sisk was wounded in the neck and bleeding to death. "The rest of the company was pinned down, and everybody looked at each other. Something said, 'Jerry, you need to do something.'" He believes the prompting came "by the good grace of the Lord... so I crawled."

Hoping to elude snipers, Murray left part of his ammunition behind and crawled—exposed to enemy fire—to Sisk in the trench. He found a towel and pressed it up against Sisk's wound "to try to cut the bleeding off, or at least pressurize it or stabilize the bleeding," Murray recalls. He kept the towel on Sisk's neck for "many, many, many" precious minutes, until the Navy corpsman could arrive to give medical treatment.

"We finally got a helicopter in, and we got him on the helicopter under fire," Murray says. He found out a few days later that Sisk had survived. "He thanked me personally in June of '68, when he came back to the field. He said, 'Lance Corporal Murray, if it wasn't for you, I wouldn't be out here today.'"

Murray received a Silver Star for his actions that day. "Knowing the man is still on this planet today, that's the biggest Silver Star that can be given," he says. "Medals to me are fine and dandy, but what's more important is knowing I made a difference for somebody's life." Murray unknowingly made another difference the day he left South Vietnam. A Marine who had been in country just six months—and was scared he wouldn't make it home—was watching him depart the field for the rear. "He later told me, 'Jerry, when I saw your smiling face in the window of that helicopter, you waving, that gave me all the hope I needed of getting out of Vietnam alive.'"

As part of the 2nd Battalion, 4th Marines, also known as "the Magnificent Bastards," Murray survived the three-day Battle of Dai Do, considered one of the fiercest engagements of the Vietnam War. He received a second Silver Star for saving the lives of six Marines, including his battalion commander, during that battle.

— **Jerry Murray**
U.S. Marine Corps, Lance Corporal, 2nd Battalion, 4th Marines, 3rd Marine Division
Vietnam: 1968–69

Navy nurse Lieutenant Commander Dorothy Ryan checks the medical chart of Marine Corporal Roy Hadaway aboard the hospital ship USS *Repose* (AH-16) off South Vietnam. *Photo courtesy of NARA.*

Jerry Murray (right) and Gunny Brando of the 2nd Battalion (aka "the Magnificent Bastards"). *Photo courtesy of Jerry Murray.*

Soldiers of the ARVN 1st Division move to a helicopter pick-up point as part of Operation Lam Son.
Photo courtesy of NARA.

A Matter of Perspective

"Lieutenant, you look like you're down in the dumps today," said the Soldier in the hospital bed next to Leland Burgess. "What's the problem?" Burgess didn't really feel like talking. Yeah, he was down in the dumps all right. He'd just had surgery on his wounded arm and his doctors weren't sure he'd ever be able to use it again.

The week before, Burgess's H-23 helicopter had been shot down on a morning that had begun well enough: a dawn patrol up at the Cambodian border, a refueling and a donut, then some low-level hedgehopping looking for Viet Cong. It was the second day of the Tet Offensive of 1968.

"We'll just do one more pass," he radioed back to the base. "Save lunch for us." He and his crew had spotted a large Viet Cong bunker complex, had called in unsuccessfully for Air Force backup, and were now trying to attack the bunkers on their own. Fifteen feet from the ground, they were hit by enemy fire, and the helicopter crashed in the middle of the Viet Cong camp.

Burgess and his two crew members took refuge in a shallow bomb crater, then made a run for it across a rice paddy, this time hiding behind a dike. Burgess, wounded in the crash, was losing a lot of blood. It was the trail of blood that helped another Army pilot spot the trio; the pilot disobeyed orders and landed his UH-1 in enemy territory to make the rescue.

"That's the best feeling in the world, when you're still alive and you're on the floor of that Huey," Burgess recalled. "It doesn't matter if you're wounded. You're out of there and you're still alive."

And now here he was, lying in a hospital room in Japan, feeling sorry for himself. "Well, I sure hope you get some good news today, like I just got," said the Soldier in the next bed. "I wasn't really that interested in his good news," Burgess admitted later. "I was thinking about myself. But I said, 'What's your good news?'"

"Well, sir," said the Soldier, "I was riding on a track that hit an anti-tank mine and it blew up the track and I lost my right leg above the knee, and my left arm above the elbow, and the left leg just below the knee. But they just came through this morning and told me my knee's healing just fine. They told me they can fit me with artificial limbs, and in six months I'll be back home, dancing with the girls."

Burgess said he remembered that Soldier's gratitude for the rest of his life.

— Leland Heywood Burgess Jr.
U.S. Army, Lieutenant, helicopter pilot, 25th Infantry Division
Vietnam 1967–68

A medic from the 1st Battalion, 16th Infantry searches the sky for a helicopter to evacuate wounded buddy following air assault. *Photo courtesy of NARA.*

The Supply Hooch

As a supply specialist at the 25th Infantry Division's base camp at Cu Chi in 1968, Lynn Richardson was responsible for the motor pool and the supply hooch.* No matter what, his supply sergeant told him sharply, the trucks had to keep moving.

"My job in Vietnam," explains Lynn Richardson, "like Radar O'Reilly's,** was to make sure the supplies needed for the motor pool were always available, day or night."

The job could be very demanding, especially when the supplies he needed exceeded the "SOP"—the Standard Operating Procedures that dictated how much of any given item he could order. Richardson became adept at trading with other units to get extra items, especially during the monsoon season when truck repairs became more frequent.

Richardson's creativity came in handy one day, when he learned on short notice about an upcoming high-priority IG (Inspector General) inspection.*** "I had all this extra stuff that I wasn't supposed to have," he explains, "and I couldn't get rid of it, but I wasn't going to let them know I had it. We needed every last bit of that stuff."

Then Richardson had an idea. He dug a 4x4x4-foot hole in the ground outside the hooch and buried the excess supplies in it. He then laid plywood over the hole and covered it with dirt. But it occurred to him that the freshly dug dirt would draw attention, so he quickly collected some rocks, painted them white, and placed them in a ring around the hole.

Sure enough, the hole raised eyebrows during the inspection. Thinking quickly, Richardson explained that he'd asked his mother to send flower seeds so he could make the place look decent. The inspectors were pleased with Richardson's fine idea and congratulated him for it. Of course, as soon as they left, Richardson went back to work, digging everything up to restock the hooch.

— **Lynn Richardson**
   U.S. Army, Specialist 5, supply, 125th Signal Battalion, 25th Infantry Division
   Vietnam: 1968–69

*G.I. slang for a temporary place to stay or keep supplies (hut, rough shelter, tent, etc.).

**Radar O'Reilly, a character on the hit TV show M*A*S*H, was company clerk at a MASH unit during the Korean War.

***An IG inspection examined every aspect of an Army unit's administration and logistics. Failure to pass could result in a commander's relief from duty.

They're Not Going to Leave His Body

When Lee Ewing got back from Vietnam, he went to his local five-and-dime for a sandwich. Sure, the guy behind the counter said, you can buy the sandwich—but you can't eat it here. *You can go and kill for this country and damn near die for this country,* Ewing thought, *but if you're a black man you can't sit down and eat a sandwich?* It was 1968.

Ewing nearly died in South Vietnam trying to retrieve the body of his buddy Charles Sheehan from a bridge. Sheehan was white, and just 17 years old when Ewing first met him in heavy equipment school after Marine boot camp. "He kind of looked up to me as the big brother, and I took a liking to him," Ewing remembers. In Vietnam, both men built and repaired the bridges that helped keep Highway 1 open. Ewing was a crane operator—an unarmored, easy target for enemy fire.

The day his buddy Sheehan was killed, their unit was making its way across a bridge and was attacked. The fighting was so heavy that Ewing's lieutenant said they'd just have to leave Sheehan's body there on the bridge.

"And I said, 'Oh, no, they're not going to leave his body.' And my lieutenant said, 'Ewing, you're crazy. We can't afford to lose you.' And I said, 'Man, I'm going to get him, he's my friend.'" Ewing was shot before he got to Sheehan. He spent one month in a hospital in Japan, then transferred to a naval hospital in Queens, New York.

"I remember after getting out of the hospital, coming home and my mom happy to see me and she said, 'Well, what happened over there?' And I started to tell her a couple of things about what had gone on in 'Nam, and she said, 'I can't take it, don't tell me no more.' And I knew then, if my mother who loved me and brought me into this world could not stand to hear it, then it wasn't supposed to be talked about. And I did not utter a word about the Vietnam War for the next 20 years."

By then, Ewing was addicted to drugs and alcohol. "I had hit rock bottom, physically, spiritually, and emotionally," he says. An art therapist at the VA Hospital eventually got him to talk about the war. Now he's 25 years sober and works as a state service officer for the VA Medical Center in Louisville, Kentucky, helping other disabled veterans get the help they need.

— **Lee S. Ewing**
   U.S. Marine Corps, Corporal, 3rd Battalion, 7th Engineers
   Vietnam: 1967–68

Leaving the Body Bag Behind

The first time in Vietnam that Dennis Reeves was issued a body bag,* he thought, *What's this?* "It's what you're gonna die in," he was told.

"Whenever we knew there was going to be a heavy battle, they would issue body bags to us," he recalls. The Soldiers who survived those battles put the body bags to good use by sleeping in them. "In the mountains it was so heavily dewed, and with the moisture at night, the leeches were everywhere," Reeves says. "For some reason they didn't like the body bags. It was the texture. A little glossy, and it would reflect the moisture."

Reeves came close to going home in one of those bags, but a buddy, Ralph Clarence Franklin, helped him cheat death. "We were setting our firing positions up, everything like clockwork," he recalls. "And all of a sudden, the first rocket-propelled grenade (RPG) came in, just da, da, da, bang! And then everything lit up. And we began firing back and forth." During the battle, he says, "Frank got up, came across the little open area, grabbed me by the collar and actually lifted me up and slung me behind this little berm.** Three seconds later, an RPG hit right there where I had been sitting. I would have been dead. He saved me."

Later, when Reeves asked why Frank had suddenly moved him to the berm, "All he would give me was that big-ass grin."

Reeves took part in much of the war's heavy combat, from July 1968 through September 1969. "We had a saying: You either died in your first month or your last month. The first month, you had no knowledge. The last month, you're overprotecting yourself. As you get to be an old-timer, your foxhole gets deeper… You learn the sights, the sounds, the noise of battle. And you see yourself transforming from that country boy from Iowa into a different person."

When the day came that Reeves's lieutenant pointed to a helicopter off in the distance and said, "That's your bird," he realized he was going home. "The emotion that drains from you is so immense." His first thought was for his men, "Because you led them through all this. But deep down, you know they're gonna make it, because you taught them, you showed them all the tricks of the trade."

Reeves jumped on the chopper, gave his men the peace sign, and kept watching them as they got smaller and smaller. "It's final now—you survived it," he says. "You did it. You're magna cum laude of the University of Vietnam. Some didn't graduate. Some did pretty damn good."

— **Dennis L. Reeves**
   U.S. Army, Sergeant, 501st Infantry (Airborne), 101st Airborne Division
   Vietnam: 1968–69

*A heavy black rubber bag approximately 36"x 90," with handles and a zipper, used to remove human remains from the battlefield.

**A berm is an earthen embankment.

Trading Nylons and Heels for Fatigues

When she first tried to enlist in the Marines right after high school, Mary Glaudel-DeZurik was told she had missed the minimum height requirement by half an inch. She was sent home to do "stretching exercises," and when she came back they measured her with shoes on. "I think they just wanted to know if I was dedicated enough to go through with this," she says.

For the first five months of her tour in South Vietnam, "incoming" meant maps and reports. Glaudel-DeZurik worked in mail and document distribution at Military Assistance Command, Vietnam (MACV) headquarters near Saigon, keeping track of the thousands of documents arriving every month.

And then came the night of January 31, 1968. She remembers sitting in a co-worker's room in Saigon watching the fireworks that kicked off Tet, the Vietnamese lunar new year. When she woke up the next morning, she learned that Saigon and a hundred other cities and towns throughout South Vietnam had been simultaneously attacked by round after round of incoming enemy rockets, mortar fire, and infantry.

After that, everything was different, she says. The women who worked in the MACV offices were told to trade in their usual uniform skirts, pumps, and nylons for men's fatigues and canvas jungle boots. That proved difficult for Glaudel-DeZurik, who wasn't even five feet tall; eventually, she had to wear a South Vietnamese Army uniform. But she was happy to be out of those sticky nylons and uncomfortable high heels, which of course had to be spit-shined every day.

After the Tet Offensive, more reports and maps started pouring in to MACV headquarters. "I didn't have a day off for three months," she says. They worked 11 hours a day, sometimes sleeping in the office. "You were always working," she remembers. "But the one thing I always thought was, *No matter how hard it is for me, the guys out in the field have it a thousand times harder*. You recognized that what you think is hard isn't hard. You kept it in perspective."

— **Mary Glaudel-DeZurik**
   U.S. Marine Corps, Sergeant,
   Military Assistance Command, Vietnam (MACV)
   Vietnam: 1967–68

Marines bring captured Viet Cong prisoner into a collection area. *Photo courtesy of NARA.*

Going Commando

Sometimes, finding something to joke about helped ease the tension of being in constant danger.

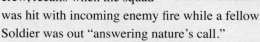

David Weinfurter, a squad leader of an 81mm mortar crew, recalls when the squad was hit with incoming enemy fire while a fellow Soldier was out "answering nature's call."

"So the guy jumped up and ran in, and his pants stayed out there," Weinfurter says. "The whole firefight went on, with him with no pants on. We didn't wear underwear over there because it's so hot all you get is a rash. So you just wore your jungle fatigues."

When the fight was over, the Soldier went back to retrieve his pants. "He never did live that one down," Weinfurter says.

— **David Weinfurter**
U.S. Army, Specialist 5, mortar squad leader,
82nd Airborne Division
Vietnam: 1968–70

Don't Shake Hands

On Christmas Day in 1968, Capt. Mel Chatman stood in an isolated field in enemy territory with two other unarmed U.S. Army officers. They were there for a meeting with a small contingent of Viet Cong —perhaps the only meeting of its kind during the Vietnam War.

The topic was the fate of several U.S. Soldiers held prisoner in South Vietnam by the Viet Cong. Chatman figures that the Viet Cong requested the unexpected meeting as a propaganda effort—trying to make the guerillas look more "humanitarian" and the South Vietnamese Army more inconsequential. Chatman, who had taught himself Vietnamese, was the assistant interpreter.

"We were warned the VC could use the meeting to make it appear the U.S. was negotiating with them without U.S. allies, the South Vietnamese," Chatman remembers. So he and the two other U.S. officers—Lt. Col. John Gibney and chief interpreter Major Jean Andre Sauvageot—were advised to avoid any "friendly gestures." Don't smile, their superior told them. Don't shake hands. Don't sit down in a relaxed manner. In fact, Chatman and the others never sat down at all, although the Viet Cong officials provided stools.

The meeting was unproductive and eventually the Viet Cong walked out on the Americans. But a week later, on New Year's Day, the three men met again with the Viet Cong. This time reporters were present, and three American POWs—Specialist 4 Thomas Jones, Specialist 4 James Brigham, and PFC Donald Smith—were led out of the surrounding forest and handed over to the American delegation. The three men were free to board the waiting U.S. Army helicopter. According to an Associated Press reporter who witnessed the historic event, Jones patted the insignia on the nose of the helicopter and rested his head against it.

Chatman continued to serve with the Army in South Vietnam until discharged in 1973. He immediately joined the U.S. Agency for International Development and remained in South Vietnam until its collapse in 1975. He was there in the last frantic weeks as the South fell to the Communists and assisted in the evacuation of Americans and Vietnamese civilians from Da Nang. Then in the last desperate hours of April 30, 1975, he did his best to help those fleeing Saigon as it fell to the Communists.

With Tan Sanh Nhut Air Base closed by enemy fire, and not enough time or helicopters to carry the number of people trying to flee Saigon, Chatman coordinated what he calls a "last-stab" evacuation by barge. As Saigon was falling, he successfully ushered Vietnamese onto buses that took them to the city docks. Chatman and another USAID officer had gathered several barges there to ferry the fleeing refugees out to sea—where U.S. Navy ships waited to carry them to freedom.

— **Mel Chatman**
U.S. Army, Captain, infantry, 199th Infantry Brigade (Light);
MACV staff; Allied Forces liaison officer; USAID officer
Vietnam: 1967–72

Survival

After arriving in Vietnam in 1968, it didn't take James McLain long to realize he would probably either get wounded or killed. The odds didn't seem to be in his favor.

His first night in the field, the North Vietnamese killed seven members of his platoon. The third night, his unit was hit with mortar rounds, and a scout dog handler and his dog were killed. The day after that, a sniper killed two sergeants and a medic.

A helicopter brought in replacements, and one was shot as he got off the helicopter. "They loaded him back on the same helicopter he got off of," McLain says. "I don't know if he survived or not."

For those first few nights, he didn't sleep at all. "When it was somebody's turn to replace me on guard, I was afraid they'd go to sleep," he explains. "I was a wreck for about three days."

So he resigned himself to either getting killed or shot. While anticipating a battle, he would feel "scared to death, walking on pins and needles." But once it started, "I don't know if it was adrenaline that kicked in, but it took the fear right out of me. I'd do some crazy stuff." And every night, McLain prayed to "thank the Lord for letting me live that day."

He managed to make ten months before getting shot, the day before he was to leave on R&R. They were patrolling in the A Shau Valley near the Laotian border, and McLain's group was told to go down and around a hill where the enemy had been seen running. McLain argued against it, thinking it was a trap. He was right.

When his team started down the hill and began crossing a creek, "they opened up on us. Boy, I mean, they were just cutting us in two."

McLain turned and saw one of his men just five feet behind him, shot in the chest and chin. He fired back—so did the three or four men still able to fight. They called for help and tried to get a wounded machine gunner back up the hill, when they were fired on again. "They caught me with slugs and hit me twice in the leg. I never felt such pain in all my life."

He spent two weeks in the Army hospital in Japan, then 11 months at a hospital at Fort Gordon, Georgia, where he was eventually able to walk with a brace. He also received a Silver Star for "gallantry in action." Instead of being upset with the news that he had been wounded, his mother rejoiced that he was alive and out of Vietnam.

"The first time she felt so good in her life, she said, was when she got that telegram saying I was wounded in action. I was in serious condition, but expected to survive. It was the first time she went shopping in a year."

— **James Hickman McLain**
  U.S. Army, Specialist 4, 2nd Battalion, 501st Infantry (Airborne),
  101st Airborne Division
  Vietnam: 1968–70

Teaching Vietnam

Nine days after getting to Vietnam, platoon commander Bob Lorish got word that his buddy Carl Hetrick—a small town boy like himself, who had gone through basic training with him—had been killed when he tripped a booby trap. Lorish was upset by the news, but knew he didn't have the luxury of dwelling on it. "You just trucked on," he remembers. "You just bottled it up and moved on." Or as the grunts used to say, "it don't mean nothin'," a phrase usually uttered with equal measures of bravado and fatalism.

Wanting to explain what the war was really like for servicemen and women in Vietnam is what propelled Lorish, 45 years later, to design a course he now teaches, along with Army vet Heyward Macdonald, at the Osher Lifelong Learning Institute at the University of Virginia.

Most of the people who take the class lived through the 1960s, but had no idea what those who served in Vietnam went through. They hadn't necessarily been anti-war, Lorish says, just indifferent.

The class covers the history and culture of Vietnam and the misery and tedium of the war, as well as an aftermath that psychiatrists now call "moral injury"—the kind of psychic harm that results, for example, when Soldiers who have been taught their whole lives "thou shalt not kill" are suddenly ordered "thou shalt kill."

Lorish hopes his course about Vietnam will help people understand the war—to bring closure and empathy, as well as insight into future American involvement abroad.

"Our citizens know so little about the Vietnam War—particularly young people," he says. "A teacher told me that in her high school American history class, a student asked, 'Was the Vietnam War before or after the Civil War?'"

— **Bob Lorish**
  U.S. Marine Corps, 2nd Lieutenant, 3rd Battalion,
  1st Marines, 1st Marine Division
  Vietnam: 1968–69

James L. Morrison, U.S. Army medic, waits with a wounded Soldier for a hoist to be raised to the Huey helicopter for a medevac.
Photo courtesy of NARA.

A flame-throwing "Zippo" boat of the Army-Navy Mobile Riverine Force burns away brush on the bank of a river in the Mekong Delta to deny the enemy concealment. *Photo courtesy of DoD.*

The Home Front

Carol Deom served at the stateside end of the war, cataloguing the body bags that were shipped back from South Vietnam on their way to loved ones across the United States. She was 17 years old, gung-ho about being a Marine, and she just did her job: the dead arrived in dry ice, she checked them off, then wheeled them into a freezer. "There was no ceremony," she says.

She thinks her ability to be so matter-of-fact about it then was her way of coping, of "not wanting to put two and two together." In her mind, "they weren't young men, they were just bodies." It was a few months later, she says, when a good friend came back from the Vietnam War missing both legs, that she started having nightmares.

Deom was stationed at Marine Corps Air Station Cherry Point, which provided combat-ready personnel for 2nd Marine Aircraft Wing units in South Vietnam. Troops were continuously arriving and departing. As Deom watched the departing troops from her vantage point in Air Freight, it seemed that "every one of them was going over there to win the war; every one of them was going to do it single-handedly. They were so young." By contrast, she says, the Marines coming back, "were old men. They were totally different."

On trips back home to Indiana, she was required to wear her dress uniform. At the airport there would be protestors, some of them also wearing pieces of military uniforms, but decorated with peace signs or an American flag tied around their waist. "I kind of just slithered through the crowd and hoped not to be noticed," she says, but now wishes she had publically revealed her pride in her service. "No matter if the war was right or wrong," she says, "you went in to serve your country in the best capacity you could." Although for years she wondered if spending the war armed with a clipboard counted as "serving," she now realizes "what I did also helped."

— Carol Deom
U.S. Marine Corps, Lance Corporal, Air Freight
North Carolina: 1968–70

Hawks and Doves Together

One day when Thomas Fox was in Basic Training at Fort Benning, Georgia, his unit was scheduled to practice a helicopter air assault, ferrying Soldiers into combat. As it turned out, all the helicopters were being used on the set of the John Wayne Vietnam War movie *The Green Berets*, which was being filmed on the base.

"We had to do our 'helicopter assault' in two-and-a-half-ton trucks," Fox recalls. "We were to drive into the landing zone and get out of the trucks as if they were helicopters."

Fox describes himself as a "reluctant warrior." From the start, he had been against the war—and that never changed, even after graduating from college and ROTC with an Army Reserve lieutenant's commission, and during graduate studies for a PhD in History at Vanderbilt University. But when Fox was called to active duty, he had a service obligation, so he went. Assigned to the Photo Interpretation Section of the 199th Infantry Brigade's Military Intelligence detachment, he fought what he called "a kind of 8-to-5 war."

"I was one of those rear-echelon characters that the grunts in the field often hated, and I can understand why they'd hate me," he says. "It was, on the whole, pretty easy duty."

In his off-hours, Fox became the brigade Education Officer and set up a center where G.I.s could take night classes and earn their GED certificate. He also taught college-level lectures on 19th-century history.

By the time he arrived in Vietnam, public sentiment about the war was fractured and intense. And that split was reflected in Fox's unit, where the "hawks" and the "doves" often argued about whether the United States should be in Vietnam. The difference between those in-country debates and the ones stateside, says Fox, was that "we all had the basic Soldier's attitude that, 'Well, it doesn't matter much if you're for or against the war—here we all are together.'"

And there was another thing they all agreed on: when *The Green Berets* was screened for the G.I.s, Fox says, "I have to tell you, even the hawks thought it was a stupid movie and made fun of it."

— George Thomas Fox
U.S. Army, Captain, 179th Military Intelligence Detachment, 199th Infantry Brigade (Light)
Vietnam: 1968–69

A Marine runs for the cover of a wall to join other Marines during street fighting in Operation Hue City. *Photo courtesy of NARA.*

Is It Going to Be My Day to Come Back?

When Thomas H. Hodge finished boot camp as a 19-year-old Marine, he was told he'd be driving a truck over in Vietnam. "I thought that might not be so bad of a job," remembers Hodge, "until my drill instructor told me the life expectancy for a truck driver was three days."

Hodge was assigned to Dong Ha Combat Base on the Demilitarized Zone (DMZ)* where North and South Vietnam were divided. It was also called the Dust Bowl, because Agent Orange had killed all the foliage. The summer temperature reached as high as 120° F and the humidity was often above 90%.

Hodge quickly learned that the names of days were meaningless. "We never kept track of the day," he explains. "It was just everyday survival. Once you hit Vietnam, you were guaranteed one meal a day, and that was it. And one hour of sleep. Sometimes we'd go for two or three days without that. If you've got Charlie shooting at you, you don't worry about sleep."

A truck driver's day started early. Hodge got up, checked out his truck, then headed over to the staging area, where the convoys lined up. Made up of 30 to 40 vehicles, the convoys included tanks stationed at the beginning, middle, and end, as well as AH-1 Cobra helicopter gunships flying overhead for protection. The convoys carried essential supplies—food, fuel, medical equipment, and explosives—and were the lifeblood for the troops. That's why convoys were such high-level targets for the enemy.

"It's like when you go to an amusement park, and you have the ducks running along in a line and you're shooting at them," Thomas explains. "That's about what it's like [being] a truck driver in Vietnam."

At the staging area, vehicles lined up, then waited for the minesweepers to "run the road" first. "Usually you don't find mines then," Hodge notes. "It's usually when we came back that we would find mines." After the convoy left, he explains, the enemy would bury pressure-detonated mines in the road. "A jeep might roll over this mine and it won't go off, but a truck carrying ammunition weighs more than two or three tons, so it would set this mine off and blow the truck up."

Hodge learned to do his job in what he describes as "a cold mind," Hodge says. "When you see buddies getting shot and guys flipping out and the enemy dying and bodies laying here and there, you don't have much choice. Your mind starts to get cold. It's something you don't learn from school; it's something that you get into there. Every time you run on a convoy, you always say, 'Is it going to be my day to come back?'"

— **Thomas H. Hodge**
  U.S. Marine Corps, Sergeant, truck driver, Company B, Force Logistics Command,
  3rd Marine Division
  Vietnam: 1968–71

*A narrow strip of land along the 17th Parallel established by the 1954 Geneva Accords that separated North and South Vietnam. Both sides agreed not to use it for military purposes, but North Vietnam violated it regularly.

7th Motor Transport Battalion (Flying "A" Roughriders) convoy from Da Nang to An Hoa. *Photo courtesy of NARA.*

Between Me and the Bullets

Scout helicopter pilot Robert V. Mitchell joked a little with the rookie observer flying with him, who asked why Mitchell seemed to always be making left turns. He grinned and said, "Well, naturally to keep you between me and the bullets." When the young Soldier looked incredulous, Mitchell explained, "Just kidding! You're the observer. For you to do your job, I need to put you in a position where you can see. You're on the left side, so that's why I always make left turns."

At the time, they were flying overhead of an infantry unit near the Laotian border. The unit's point man was ambushed and shot, and Mitchell dropped down low to scout the enemy's location.

Suddenly, automatic weapons fire hit the cockpit.

"And this kid who I've just told that I was keeping him between me and the bullets was shot through the legs three times, three places," Mitchell recalls. "He returned fire as I'm turning the aircraft to escape, and I'm calling that we're hit, we're receiving fire, and then I was hit in the left leg."

The helicopter's engine was hit, and they began going down. But since the chopper hadn't completely died, Mitchell decided to try for the closest landing zone, LZ Lash, about four miles away. "The aircraft had been shot through the fuel cell, so you could smell jet fuel," Mitchell says. "As for my observer, one of the rounds had hit his femoral artery, so he was bleeding extremely bad. He was already going into shock. I grabbed the first aid kit, handed it to him and said, 'You need to stop the bleeding on your leg.' He had his fists doubled up between his knees and was just rocking back and forth. Blood was everywhere, because it was his artery."

Mitchell was going so fast that he almost overshot the landing zone, even with his damaged engine. The wounded observer unbuckled and rolled out of the aircraft even as it was sliding along the ground.

"I could see the medics running down the hill with their aid bags in the air," Mitchell remembers. "I jumped out and ran around the aircraft. The medics were already working on him."

He ran back to the aircraft and got the first aid kit and started helping until he felt someone grab him by the shoulder. A voice said, "Son, my men will get it." It was the battalion commander at the firebase, who had noticed that Mitchell was limping. Mitchell then remembered that he, too, had been shot.

While they stabilized the observer and waited for a medevac transport, Mitchell assessed the chopper's damage. "There was blood all the way back to the tail rotor and down the side," he says. Miraculously, the rookie observer survived.

Mitchell had already been wounded once before. After recovering from this second wound, his tour was up, but he stayed in the Army almost 30 years. "The reason I stayed is because of what I found in the Army," Mitchell says. "I found guys who were willing to die for me, and I was willing to die for them. I mean, a voice on the radio calls, 'I need help,' and we'd drop whatever we were doing and we'd do whatever it took to save them. Sometimes we didn't save them—but they knew we did our best. And that's the kind of camaraderie I found throughout the rest of the time in the Army."

— Robert V. Mitchell Jr.
U.S. Army, Chief Warrant Officer 2, helicopter pilot, Troop C, 2nd Squadron, 17th Cavalry, 101st Airborne Division
Vietnam: 1968–69

Napalm bombs explode on Viet Cong targets.
Photo courtesy of NARA.

The Fire Below

In the wee hours of a September night in 1968, an alert sounded and fighter pilot Al Gardner jumped into action. Like all fighter pilots on call, Gardner slept in his flight suit. He zipped up his boots and in a few minutes was airborne in his F-100 Super Sabre.

It was only then that he learned what his mission was. The North Vietnamese Army had attacked an American infantry platoon during the night and was in a desperate situation, surrounded and outnumbered. Gardner was asked to drop napalm* nearly on top of the platoon while the men submerged themselves in the water of a rice paddy. In other words, the only way they would make it out alive was to face possible annihilation.

It was dark when Gardner's F-100 reached the battle, and guided by a white phosphorous rocket from an air forward controller who was in contact with a controller on the ground, Gardner carefully released the napalm bomb to hit the enemy but not the Americans. He'd been dropping napalm for months, so he knew exactly how to hit a target. He had become so precise, he says, that he could drop a napalm container not just on the right building but actually in the doorway of that building.

So he wasn't nervous or afraid. "You don't become a fighter pilot unless you can handle that kind of stuff," he says. During the 240 missions he flew in Vietnam, Gardner's plane was hit twice and he had previously survived a mid-air collision. "You have to have nerves of steel," he says.

He found out later that the napalm he dropped that day killed at least 25 North Vietnamese soldiers. The ground controller radioed that Gardner had executed the napalm drop "perfectly." When Gardner asked if the U.S. troops had survived, the answer was "most of them."

He never found out exact numbers of Americans saved that day—the war was too chaotic for that kind of follow-up. A fighter pilot's job was to do what was asked, and then to get up the next day and fly another mission.

— E. W. "Al" Gardner
Air National Guard, Captain, pilot, 120th Tactical Fighter Squadron
Vietnam: 1968–69

*Napalm is a mixture of gel and fuel which explodes into flames that can reach 5,000° F. It sticks to the skin and severely burns any animal or human in its path. It sucks the oxygen out of the air, causing immediate suffocation.

Gardner was a member of the 120th Tactical Fighter Squadron of the Colorado ANG, the first Air National Guard unit to be sent to Vietnam.

Four U.S. F-100 Super Sabres return to Tan Son Nhut Air Base from a combat mission. These 481st Tactical Fighter Squadron jets conducted daily strikes and provided close air support missions against Viet Cong strongholds.
Photo courtesy of NARA.

How We Passed the Time

In Hoa Lo prison—the one American POWs called "Hanoi Hilton"—the days were filled with a tedious predictability: bowls of pork fat and boiled greens shoved into the cells at 6 a.m., another bowl just like it 12 hours later, and at 11 a.m. the gong that signaled the outside world's two-hour siesta. During the quiet that followed that gong one morning, Mike Burns heard a surprising whisper.

"Cell 4, get under your door."

By then, Burns had been at Hanoi Hilton nearly six months and had heard the occasional coughing that assured him there were other Americans in other cells. But this was the first real communication with anybody other than his two cellmates. There was a two-inch space between the door and the floor, and when he peered through it he discovered the smiling face of Marine Capt. Jerry Marvel, in solitary confinement across the dim corridor.

"You need to learn the tap code," Marvel whispered to Burns. "Memorize it and get on the wall tonight." The code—five letters across, five down, a first tap signifying the horizontal row, the second tap the vertical—meant the POWs could carry on stealthy conversations, their ears pressed up against the dank walls. They could reassure each other, teach each other (math, French, Shakespeare), and break through the isolation of the interminable days.

"We talked about every single thing you can think about in your life," Burns says. They told each other poems and the plots of movies. "There was a guy who was a butcher once, before he was in the Navy, and he could draw a cow on the floor with a rock, and we talked about the pieces of meat, and spent hours discussing the pros and cons."

In one cell he occupied there was a window, and if Burns positioned himself just right he could see the top of a tree. "And one summer, I think it's 1969, I watched that tree in the morning, when light came up. I just stared at it for hours, and I could see that when the sun came up, that Hanoi sun in the summer, and just beat that tree, I could see it actually wrinkle, actually droop under that beating sun. I mean, that's how we'd pass the time."

Being a POW forced Burns to come face to face with himself, in a cell "where there's no place to hide," he says. He was determined that he would survive intact. He told himself, "I'm going to keep my head. I'm not going to come apart." Burns was released from the Hanoi Hilton on Mar. 14, 1973. Several years after he returned home, he said, "I'm very proud of my 56 months as a POW. We did not stop living in Hanoi; we just found other avenues of existence that had meaning for us."

— Michael T. Burns
U.S. Air Force, pilot, 2nd Lieutenant, 443rd Tactical Fighter Unit
Vietnam: 1968–73 (POW)

Lance Corporal Gene Davis, sniper with Company D, 1st Battalion, 5th Marines, puts his sights on an enemy position. *Photo courtesy of NARA.*

In a staged Viet Cong propaganda photo, U.S. Air Force Captain Wilmer N. Grubb appears to be given first aid while guarded by his Viet Cong captors. *Photo courtesy of NARA.*

Front view of UH-1 helicopters taking off to insert South Vietnamese troops near enemy positions. *Photo courtesy of NARA.*

The Greatest Thing that Happened to Me

On September 25, 1968, Joseph Godenzi's life changed forever. He got two letters in Vietnam that day—he still has them, now covered with dried blood. He was in the middle of reading the letters when two Soldiers came running into camp, screaming, "They're coming, they're coming!"

"We all jumped into our foxholes, our bunkers, and got ready," he remembers. "And within a couple of minutes, all hell broke loose." The enemy was everywhere, and Godenzi shot round after round from his M-16. At one point he felt like he got punched in the back, but he kept shooting. "All of a sudden, my sergeant, who was in the bunker with me, said, 'Joey, you've been hit!' I looked down and there was a big puddle of blood on the floor. I felt fine, but I must have blacked out. When I came to, my sergeant was lying across me, and the back of his head was gone. He'd been hit with an RPG."

Then another grenade landed in the bunker. Godenzi grabbed it and threw it out of the firing hole; it exploded within a few inches of his hand, blowing off his thumb and pieces of two fingers. "And then I panicked," he says. "I jumped out of the bunker and crawled through my hooch to a tree, and I laid there, screaming. All of a sudden, the tree fell on me and crushed my chest and punctured one of my lungs. I was still screaming."

The firefight went on for hours while Godenzi lay pinned under the tree. Two medics were wounded or killed while trying to help him; a third medic finally got to his side. "Because I had such serious stomach and back wounds, he couldn't give me morphine," he says. "Basically, all he could do was wrap up my hand and put bandages around my chest and back." Then the medic moved on.

"I remember looking at my hand and saying, 'Oh, my God, I'm never going to play ball again,'" Godenzi recalls. Then a monsoon rain hit, and he was drenched in the downpour while bullets continued to fly above his head. Although he was alone, his friend Ronald Stuckey was somewhere nearby. Stuckey was from Detroit, and throughout the hours-long ordeal, Stuckey called out to Godenzi, trying to stop him from going into shock. "The Tigers got into the World Series that year," Godenzi says, "and Stuckey kept saying to me, 'You're going home. You're going to go see my Tigers in the World Series.' He kept saying to me, 'You're all right. We got everything under control.'"

Eventually the firefight ended. His buddies had to use a chainsaw to cut Godenzi out of the tree and helped evacuate him to a MASH unit. He discovered later that his unit had gone back to the same area and suffered terrible losses, including his friend Stuckey.

After spending time in several hospitals, Godenzi landed at Walter Reed Hospital in Washington, D.C. "I got put in the same room with a young man who was 19 years old and had both legs gone and an arm and an eye gone," Godenzi recalls. "And this guy kept me in stitches, laughing and joking. He made me aware that I was one lucky human being because I knew I was going to walk out of that hospital and he was never going to walk again. I've never had a problem mentally, just a little bit of PTS, and I owe it all to that gentleman. He was the greatest thing that ever happened to me during my Vietnam era."

— **Joseph A. Godenzi**
U.S. Army, Specialist 4, Company C, 2nd Battalion, 35th Infantry, 4th Infantry Division
Vietnam: 1968–69

Cardboard Stationery

It was so wet during monsoon season in the jungles of South Vietnam that even if Bill Green carried paper stationery with him, it would be too damp to write on—and the high humidity would seal the envelopes before he had a chance to stuff a letter inside.

So, like a lot of grunts, Bill Green wrote home on pieces of cardboard torn from C-ration cartons. Unfolded, the pieces made a 5-by-18 inch letter, plenty long enough to tell his wife, Peggy, the news of the war. Except, of course, like a lot of grunts, he rarely talked about what was really happening.

He might say he saw a monkey. Or talk about the weather. "I never talked about combat, period," he says—unless he thought a battle might have been in the newspaper back in California. Then he would say, "If you hear about it, don't worry." Mostly he talked about how much he missed her.

He never told her about how exhausted he was all the time. How he slept four hours a night, but never four hours in a row; or how when he slept on the jungle floor, creatures were always crawling over him. He never told her about the jungle rot and the malaria.

Green tried to enlist in 1967. He was told he couldn't because he had asthma, and then was drafted two months later. He volunteered for the infantry. "If you're going to interrupt my life, make me a Soldier, not a clerk," he said.

He had married Peggy while in Hawaii on R&R. And he kept writing her letters, which she saved—every last one of them. Today they're in a box in their garage, except for one letter that's in the Veterans Memorial Building of San Ramon Valley, California. Peggy wrote him every day during the year he was in country, but he would read her letter and then burn it. "You didn't want the enemy to get the address," he says.

— Bill Green
U.S. Army, Sergeant, 198th Infantry Brigade (Light)
Vietnam: 1968

The Uniform

It's too dangerous for you to wear your uniform, Maj. Dennis Reimer was told when he landed back in California after his second tour in Vietnam. He'd just spent a year wearing his uniform in a place where he was being shot at, so this advice struck him as ironic.

"The idea that being in my home country was more dangerous than being in Vietnam didn't make sense," Reimer says. But this was 1968, when America's big cities and campuses were full of anti-war protestors who often liked harassing returning GIs. Soldiers returning to the States were warned to change into civilian clothes before they headed out in public. In those contentious days of the war, Reimer remembers, even the top brass at the Pentagon would often wear street clothes on the way to work, then change into uniforms once they were inside the building.

Reimer stayed in the Army, and during the Gulf War, he was the Army's Deputy Chief of Staff for Operations and Planning. When that war was over, he remembers standing along Constitution Avenue in Washington, D.C., during the National Victory Celebration, listening to the crowd cheer the troops.

The next day, Reimer and his wife drove by a church in suburban Virginia. "And there were two Marines out there in their uniforms, as proud as can be," he remembers. "For me, that was closure on Vietnam. The uniform was respected again."

— Dennis Reimer
U.S. Army, Major, artillery, 2nd Battalion, 4th Artillery, 9th Infantry Division
Vietnam: 1964–65 (advisor); 1967–68

General Dennis Reimer served as Chief of Staff of the U.S. Army (1995–99).

Protesters stage a peaceful sit-in at the mall entrance to the Pentagon.
Photo courtesy of NARA.

The Bad News

"Oh, you're in the Army. My daddy's in the Army," said the six-year-old boy, looking up at the solemn man on the doorstep in a military uniform. Gerald Newton swallowed hard, because he had come to tell the boy that his daddy wasn't coming home.

Stationed in New Bedford, Massachusetts, in the mid-1960s, part of Newton's assignment was notifying families when a loved one was killed in South Vietnam. "Hardest thing I've ever had to do in my life," he says. "Knock on doors. You put on a dress uniform, you go in a military sedan, and as you pull up into the neighborhood, everyone knows. There is just no doubt as to why you're there. And to have to go up and tell the parents or spouses."

In the case of the six-year-old boy, his parents were separated. The boy was listed on his father's emergency notification form as the person he wanted told in the event of his death.

Newton also arranged the military funerals and ceremonies for posthumous awards. Since he was the family's only connection to the military, many pressed him to find out how the Soldier died. Sometimes the knowledge gave them closure; other times, it only added to their sorrow. And sometimes there were no answers.

"That was the way it was handled back then," Newton explains. "Now they send generals to the homes and they have a chaplain. I used to go out alone. Back then there was no e-mail. We would get a Western Union teletype message, and they told me exactly what to say. My speech, essentially, was, 'The Secretary of Defense regrets to inform you that your son, father, brother—whatever it is—was killed in action.' And they would insert the name of the town in Vietnam. 'And the Secretary deeply regrets this loss,' and so forth. Now it's much more personalized."

Rather than continue notifying families of the deaths of their loved ones, Newton volunteered for duty in Vietnam. "Seeing these young kids getting killed and then having to notify the families really brought it home in a very unique way," he says. "I was career military. I couldn't change the government's position, but I had a responsibility to keep these youngsters alive and to do the best that I could."

In 1968, he was sent to Headquarters, MACV (Military Assistance Command, Vietnam) in the outskirts of Saigon, where he was responsible for handling everything relating to casualties, mortuary services, and prisoners of war. He worked from 7 a.m. to 7 p.m. every day of the week. Every third night, his 120-man unit, comprised mainly of the headquarters enlisted staff, spent all night guarding the defense perimeters against infiltration by Viet Cong. "They would try to get in at night, shoot up the building, shoot at the barracks, or drop a mortar or something and kill a couple of Americans," he says. "Terrorism is basically what it was."

"The most fortunate thing was that one of my platoon sergeants had served in Special Forces in South America," he continues. "Very talented guy. A real soldier's soldier. And he really helped me run that unit."

Looking back at his time in Vietnam, Newton remembers how tired he was all the time. "Working 12-hour days and doing those night patrols and things, you were just so fatigued that everything was sort of a big blur." But he preferred it to notifying families. "I always thought about that when I was in Vietnam," he says, "that I was happier working at the headquarters in Vietnam than knocking on doors in the U.S."

— **Gerald "Jerry" Newton**
U.S. Army, First Sergeant, J-1 and J-13, Military Assistance Command,
Vietnam (MACV); Company B, MACV Defense Force
Vietnam: 1968–69

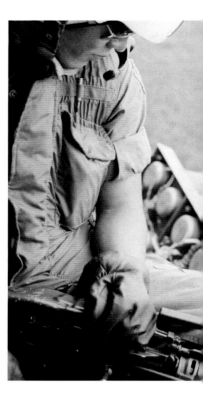

A U.S. Army Huey helicopter arrives in the hills of Kontum Province.
Photo courtesy of NARA.

U.S. Navy Seventh Fleet ships and landing craft transport Marines on shore near Da Nang. *Photo courtesy of DoD.*

Soldiers examine captured enemy documents found while on patrol. *Photo courtesy of Rocky Olson.*

Lone Survivor

Killing the enemy was a moral dilemma for Rocky Olson, who went to Vietnam after serving a two-year mission for his church. But he tried to live by the scripture in Ecclesiastes: *To everything there is a season... A time to kill, and a time to heal... A time of war, and a time of peace.*

"I was not anxious to go off to war, but I believed in fighting for my country," Olson says. On Dec. 28, 1968, his 12-man squad spent the night on top of a small hill between the Annamite Mountains and Camp Evans in northern South Vietnam. Their position overlooked some abandoned rice fields. "At sun-up we visually surveyed the old rice paddies that were now covered with waist-high elephant grass," Olson recalls. "From our vantage point, we couldn't see a thing out there that would be dangerous."

The squad spread out single-file and carefully descended into one of the rice paddies. Once the Soldiers were in the waist-high grass, the hedgerows on two sides erupted with AK-47 and machine-gun fire. Thirty North Vietnamese Army soldiers had the squad trapped in a classic L-shaped ambush.

"We had no place to hide," Olson says, "so we prayed and returned fire as best we could. The RTO [radio operator] in front of me called for help on the radio, but he was quickly struck and killed by a machine-gun burst. Within 30 seconds, I believed everyone was either dead or seriously wounded. I could hear several [Soldiers] calling for their mothers or the medic, but we had neither."

Olson kept firing his M-79 grenade launcher almost straight up in the air. "I was trying to make the grenades return to earth [close by], forcing the enemy in that hedgerow to back up away from me. Our only hope was if that radio call had made it out." Lying in the elephant grass, Olson watched green tracers zip within inches of his body, which he tried to make as small as possible. Suddenly, he heard his squad leader yell, "Rocky! Let's go get them!"

"He knew we were going to be killed within seconds, but we wanted to die fighting like Soldiers, not cowering in the grass," Olson explains. "So we jumped up and charged the nearest hedgerow, my squad leader firing his M-16 and me firing my M-79 as fast as I could put the grenade rounds in the chamber."

Two American gunships soon arrived and began to rocket and strafe the thick foliage. Another squad of Company A infantry quickly came running to the ambush site and helped force the enemy to withdraw. Once the medevacs arrived, Olson helped get the eight severely wounded and three dead Soldiers onto the helicopters, which took off quickly.

The courageous squad leader was killed an hour later, after stepping on a NVA landmine. Olson remained the sole unwounded survivor of the battle. "I cried a great deal," Olson says. "It was a tough thing to lose all my friends, my squad brothers, at one time. And valiant Soldiers deserve to be cried for."

For five years after coming home, Olson says, it was too difficult to share war experiences with friends or family. But he later wrote them in a book, *Sgt. Rock: Last Warrior Standing*, published in 2011. "Many Soldiers go to war thinking they could never kill someone," Olson says. "But the first time one of your very best friends gets shot and killed, and you can see the person who did it, you learn how to kill very quickly. The next time, you'll be the first to shoot. You won't wait for your friend to be killed. Sometimes a learning lesson is a terrible thing."

— **Rocky Olson**
U.S. Army, Sergeant, Company A, 2nd Battalion, 506th Infantry (Airborne), 101st Airborne Division
Vietnam: 1968–69

U.S. Air Force C-130 Hercules keeps its engines running for a quick takeoff while unloading supplies. *Photo courtesy of NARA.*

Finding an Audience, 40 Years Later

Every day during his year in South Vietnam, Sgt. Steven Burchik wrote a note home to his future wife, Christina. And, more often than not, he dropped in the envelope black-and-white photos or a roll of color slide film he'd shot while stationed northeast of Saigon.

Burchik and his platoon spent most of their time on patrol, trudging through rice patties and jungles, alert to the threat of ambush. With his commanding officer's permission, Burchik took his camera with him about twice a month. He took photos during rest breaks, capturing images of his buddies and the nearby villages. Eventually he had more than 4,000 photos.

When he got back to the United States, he put together slideshows for his family and friends. But it was 1969. "People were overwhelmed with the war and tired of it," he soon realized. "They didn't want to see the pictures." So he put them away.

And then one day, 40 years later, a family friend who was teaching a high school class on war and literature asked Burchik to speak to her class. Remembering the cold shoulder his photos had gotten when he first returned from South Vietnam, he hesitated at first, then gave in. And not long after that, the teacher presented him with 60 letters the students had written him. They had found his talk interesting, they said. And they thanked him.

"It was the first time anyone had said anything positive to me about the war," Burchik says.

And then more speaking requests rolled in. At one presentation, a listener asked when his book would be out. "It was something I hadn't considered," Burchik says. "But I started scanning the letters I sent home, matching them up with the photos, so it was like a chronicle of my year in Vietnam."

That 2014 memoir, *Compass and a Camera: A Year in Vietnam*, has led to gallery exhibitions and speaking engagements two or three times a week.

"For some people, it's the first time they've interacted with anyone who served," he says. "And when I show the pictures to veterans groups, it seems to provide an emotional release."

— **Steven Burchik**
 U.S. Army, Sergeant, forward observer, mortar platoon, 1st Infantry Division
 Vietnam: 1968–69

Soldiers trek through a marsh.
Photo courtesy of Steven Burchik

Injured Soldier being evacuated.
Photo courtesy of Steven Burchik

Vietnamese children. *Photo courtesy of Steven Burchik*

Dropping into a battle zone by helicopter.
Photo courtesy of Steven Burchik

A Marine M60 machine-gun crew prepares a defensive position under direction of a sergeant along a Vietnamese hillside. *Photo courtesy of NARA.*

A Camera is Worth a Thousand Words

While serving as the staff judge advocate for the 9th Infantry Division, Lt. Col. John Franklin Webb encountered cases where he felt legal technicalities should not apply. One example of this dilemma was when a combat photographer was shot while taking photos during a battle. The photographer had raised a 35mm camera up to his eye to snap a picture when a bullet hit the front of the camera lens, ricocheted off, hit the camera body, continued up under the Soldier's helmet, and dropped down the back of his neck.

"The impact of the bullet drove the camera into the photographer's face, broke his nose, and gave him a black eye, but he was otherwise uninjured," Webb says.

The impact left the camera cracked, bent, and dented. It no longer worked nor could be repaired, so the Soldier asked to keep it. After all, it had saved his life—if the camera hadn't been raised up to his face at that very second, he would have been killed. But regulations are regulations—and all unrepairable government property on record had to be turned in for salvage, where it would likely be buried or destroyed.

The Division Information Officer asked if Webb could find a legal way to give the Soldier the camera. Webb couldn't think of a way around the law, so he suggested they share the story with the division's commander, Maj. Gen. Julian Ewell.

"I told him we should write the camera off as destroyed or lost in combat and present it to the young Soldier," Webb recalls. "General Ewell went one better. He had the camera bronzed, placed on a plaque, and presented it to the young Soldier as a trophy with a nice ceremony. Ewell also personally presented him with a Purple Heart for the injury to his face sustained in combat."

Webb wonders if somewhere in an archive there's a piece of paper saying a 35mm camera was lost, or turned in and salvaged. "But more importantly," he says, "I hope that somewhere there is a man, now middle-aged, who has a battered bronzed camera that still means more to him than any other memento in his life."

—John Franklin Webb Jr.
U.S. Army, Lieutenant Colonel, staff judge advocate, 9th Infantry Division
Vietnam: 1968–69

Soldiers First, Artists Second

When Don Schol headed off to his first assignment during the Vietnam War, he packed ammunition and an M-16, just like his fellow Soldiers. But unlike his comrades, Schol also carried a camera, a sketchbook, and artist pencils. He was one of only 45 combat-trained Soldiers in the U.S. Army Vietnam Combat Artist Program, the first such program since WWII.

"It was unique," Schol says. "You had to be a combat-trained Soldier, but you also had to have credentials that verified you were an artist." Not long after he had received a master's degree in fine arts, Schol received a draft notice. He enlisted in Officer Candidate School, and after being commissioned, soon found himself in South Vietnam documenting the visual history of the war from an artist's perspective.

"Historically, there's a tradition of combat art in the United States," Schol explains, "but in Vietnam, they wanted us to look at the war in a different way. They wanted us to go into the field as Soldiers, but also to observe the war and participate and record our experiences in a more artistic manner than just a photograph."

The 45 combat artists traveled throughout South Vietnam in nine teams of five. The men carried a set of orders that gave them carte blanche to travel wherever and however they wanted. "Periodically, the colonel briefed us where particular action was taking place," Schol explains. "It was my responsibility to get my team to that location however we could, whether we flew by prop plane, helicopter, jeep, or whatever. I would announce our arrival to the commander of that particular unit and let them know what we were up to. And then we went on various operations with them."

While the troops were underway to a particular location, Schol, who was trained as a sculptor, drew landscapes and portraits to help him sculpt figures later. "When the bullets started flying, I put my sketchbook away and paid attention to what was going on," he says. "The colonel told us, 'I want you to remember you're Soldiers first, artists second.'"

Schol spent six months on active duty in South Vietnam, then an additional six months in Hawaii, working in a fully equipped art studio to create finished works of art from his sketches and photos that would become part of the National Archives War Collection. He keeps a sense of humor about where his art ended up. "When people ask me where my work is, I've always said, 'Do you remember that last scene in *Raiders of the Lost Ark*? The huge warehouse full of big crates?'"

But after five decades, they've started opening up the crates. "There's a lot of work to look at," Schol notes. "They've never had anybody who wanted to open that up and go through it. Now they're trying to catalog it all."

— Don Schol
U.S. Army, 1st Lieutenant, combat artist, U.S. Army, Vietnam (USARV)
Vietnam: 1967–68

Sergeant Clarence Weitzel, squad leader, Company D, 2nd Battalion, 503rd Infantry (Airborne), 173rd Airborne Brigade, keeps a sharp eye past the perimeter with his M60 machine gun in preparation for the final assault on Hill 875 in South Vietnam.
Photo courtesy of NARA.

He Gave Me This Life

The new guy that day was named Fous, and John Sharp sized him up immediately as somebody who was paying attention and listening to learn the tricks for survival. When they advanced across an area of open, flat rice paddies in the Mekong River Delta—a place where new guys who aren't paying attention might bunch up and turn all of them into an easy target—Jim Fous kept his spacing. He was a listener. He was a skinny, nice kid who sounded like he was from somewhere in the Midwest.

It was Fous's first operation since arriving in South Vietnam two weeks earlier. Sharp had been there seven months, already a veteran old-timer at age 21. That May night in 1968, with darkness closing in, Sharp's company set up a night defensive perimeter. Sharp told Fous to take the first lookout duty so that Sharp and other Soldiers could take shifts later in the night, when the Viet Cong were more likely to attack.

"I awoke to Fous shooting his M-16," Sharp recalls. "I remember sitting straight up and I was facing the same way he was shooting and I saw a gook going left to right, across the tree line." Sharp began scrambling for his own gun when all of a sudden he heard Fous yell, "Grenade!"

Sharp dove to the ground to get as flat as he could and then heard the grenade go off, bracing for the pain that would follow. But when he opened his eyes, he was alive and Fous was clutching his mortally wounded chest. The new guy had died saving the lives of three Soldiers he barely knew.

Before he went to Vietnam, all Sharp wanted to do was race his '57 Thunderbird and party on the weekends. His friends told him later they had voted him "most likely to be arrested." But the John Sharp who went to war didn't come back. "Vietnam totally turned me upside down," he says.

When he got home he became a policeman. In 1973—it took him five years to get up the courage—he called Fous's mother and, sobbing into the phone, talked to her about her son's selfless courage. And not long ago, when he found out another grandchild was on the way, he closed his eyes, and, as always, thanked Jim Fous.

"He gave me this life."

— John Sharp
U.S. Army, Sergeant, Company E, 4th Battalion, 47th Infantry, 9th Infantry Division Vietnam: 1967–68

For his "gallantry and intrepidity in action," James W. Fous was posthumously awarded the Medal of Honor in 1970.

Hunger Pangs

In 1966, faced with the prospect of being drafted and sent to Vietnam, Terry McNamara signed up with the Idaho National Guard. Ironically, less than two years later Idaho became one of only eight states from which Army Guard units were called up, and McNamara deployed to Vietnam with the 116th Combat Engineers.

The 116th consisted of men in their early 20s, most of them farmers or loggers, and many of them Eagle Scouts and hunters. These were men who knew how to make use of anything at hand to accomplish the mission. "We weren't city boys," explains McNamara. "We knew how to get our hands dirty when we needed to."

Those Idaho skills came in handy in South Vietnam's highlands where the unit was based. One hot afternoon, several engineers from the 116th were riding back to camp in their truck when they saw some of the Montagnard village women who lived with their families near the engineers' compound. The Montagnard—French for "mountain people"—were one of several indigenous ethnic groups that had lived in Vietnam for centuries. Many of the "Yards," as the Americans referred to them, had associated with the French during the colonial period. But when the French withdrew from Vietnam, the ethnic Vietnamese ostracized the "Yards" as primitives.

As the engineers approached, they saw that the women were all carrying woven baskets slung on their backs, and their teeth were blackened from years of chewing the betel nuts that staved off hunger pangs. "As we got closer to them," McNamara recalls, "we could smell a terrible, putrid odor. Then we could see a slimy substance dripping from one of their baskets." The men soon realized that what they smelled was the rotting carcass of a pig.

"We thought we had seen a lot of things in our lives, but this was just plain disgusting," McNamara says. The Guardsmen realized that the women were taking the carcass back to the village, in hopes of scavenging some usable portion from it. Only then did the men fully understand how impoverished the villagers were.

From that day on, whenever the men had to kill an animal that had become entangled in the compound's perimeter fence, they would take it to the village. The first time the engineers killed a deer, they gutted it before taking it to the village. The villagers gratefully welcomed the gift, but were disappointed when they discovered the entrails were missing—like many native people, they used every part of an animal. After that, the engineers would deliver the carcasses intact to the village, and were gratified to know there would be less hunger among the tribe that night. "They deserved our help, and we were glad to give it," McNamara says.

— Terry McNamara
Idaho National Guard, Specialist 4, 116th Combat Engineers Vietnam: 1968–69

Two Soldiers return fire with M79 grenade launchers from an infantry firebase. *Photo courtesy of NARA.*

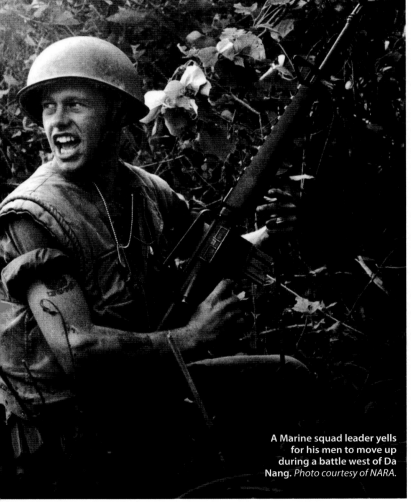

A Marine squad leader yells for his men to move up during a battle west of Da Nang. *Photo courtesy of NARA.*

The Grenade

On a May morning in 1968, Navy corpsman Donald Ballard was moving with Company M, 3rd Battalion, 4th Marines in Quang Tri Province of South Vietnam when North Vietnamese soldiers sprang an ambush. Kneeling over a wounded Marine, Ballard looked up with horror to see an enemy grenade rolling down the hill toward them.

You don't have too many choices with an approaching grenade, he points out. "It was acceptable but inappropriate to throw a dead body on it, something to absorb the blast. But I didn't have any volunteers, and nobody wanted to play dead." And the wounded Marine clearly couldn't get up and run away. That left only one choice, Ballard remembers.

"I didn't want to commit suicide. I had a wife and two kids," he says. "I had a life and I loved myself as much as I did the Marines." But he had to do something. "I thought I could absorb the blast and save their lives. I believed it was going to kill us all if something wasn't done."

Ballard was wearing a flak jacket. "I wore that jacket all the time except when I was in the shower. I even slept in it. I guess I was thinking my body would take most of the blast and save the others."

So he lunged forward and pulled the grenade underneath his chest and waited. "It seemed like an eternity," he remembers. When the grenade failed to detonate, a second instinct kicked in.

"I was lying beside one of my patients, and as I rolled up off the grenade I turned over onto him… and in one motion I slung it down the hill as I rolled." And then he began to worry: what if he'd thrown the grenade onto one of the other Marines?

But everyone was saved, and in 1970, Ballard received the Medal of Honor. He downplays his actions, though. "I didn't even think anybody saw what really happened. It didn't appear to me worthy of a general flying in and saying, 'You're a hero.' There's a lot more deserving people out there. I was just doing my job."

— **Donald Ballard**
   U.S. Navy, Corpsman 3rd Class, Company M,
   3rd Battalion, 4th Marines, 3rd Marine Division
   Vietnam: 1968

Excerpt from: Herman, Jan K. *Navy Medicine in Vietnam: Passage to Freedom to the Fall of Saigon*. The U.S. Navy and the Vietnam War. Eds. Edward J. Marolda and Sandra J. Doyle. Washington, D.C.: Naval History & Heritage Command. 2014

Two Very Different Tours of Duty

For years, Rodney Goehler couldn't even talk about what he did during his first tour in Vietnam. He served in the Studies and Observations Group (SOG), an elite unit whose missions beyond South Vietnamese borders were kept secret by the U.S. government for more than 20 years.

"We were actually going across the border into Laos as reconnaissance teams, gathering information and bringing that information back," explains Goehler, who led a 12-man team made up of three Americans and nine Chinese immigrants from Vietnam. Nobody spoke all three languages (English, Vietnamese, and Chinese), so communication involved "mostly hand and arm signals," Goehler says.

According to Goehler, SOG units lost approximately 78 percent of their Soldiers. "Unfortunately, they were [listed] as missing in action because of our assignment," he says. "When we went there, we went in plain uniforms—there weren't any markings on it. We didn't have any dog tags. We didn't have any identification. And we had foreign weapons. Nothing to tie us back to the U.S. Army."

Goehler's second tour of duty was more conventional: he was company commander of a regular infantry unit. Normal operations had his unit protecting his battalion headquarters for one week, then rotating out to the field for three weeks.

During his second tour, Goehler was injured by fragments in his left arm and hand. "We were setting up our defensive perimeter for that evening, and I had just assigned my platoons to set out ambushes, and all of a sudden the mortar rounds started coming," he recalls. "I just happened to be in the wrong place. In Vietnam, I don't think there were too many nights that went by that you didn't get mortared."

The company medic "doctor[ed] me up," he says, so Goehler stayed in the field. "The difference between my tour in '66 to '67 and the return in '68 to '69 was like day and night," he observes. "For example, when the sun went down in 1966, if somebody moved out there, you would take them under fire, no questions asked. They weren't supposed to be moving after sundown."

"When I went back in 1968," Goehler continues, "to great shock to me, I had to notify the battalion that I had movement to my front or that one of my platoon leaders had movement, whatever it might be. They in turn would have to go to the brigade, which is the next step up in the chain of command. And the brigade would have to go to the Vietnamese counter-structure and ask them for permission to say, 'Go ahead and open fire.' That's how bad it was, the difference of two years. The politics of that war became so outrageous."

— Rodney W. Goehler
U.S. Army, 2nd Lieutenant, 5th Special Forces;
Captain, company commander, 3rd Brigade, 82nd Airborne Division
Vietnam: 1966–67, 1968–69

Troops enter the rubber
plantation at Loc Ninh.
Photo courtesy of NARA.

It Mattered That I Had Been Here

Sneaking in a bottle of champagne for the wounded Soldiers in the Navy hospital in Corpus Christi sounded like the perfect way to ring in the New Year, recalls Navy nurse Gail Gutierrez. "I felt so sorry for those guys who didn't make it home for the holidays," she explains, "so I went around and gave all of them a glass. Of course, I would have been in real trouble if I had been found out, but there are some times when you just have to bend the rules."

Gutierrez joined the Navy because she wanted to do her part in the war effort. Her goal was to work on a hospital ship or in the Navy hospital in Da Nang, but the Navy required a year of experience first, so she was sent to Texas. "I was unsure of myself and not very sophisticated. I didn't really know how to talk to these young men who had come back from war," Gutierrez recalls. But she must have been a favorite, because drinking wasn't the only rule she helped her patients bend.

"The men weren't allowed to smoke," she explains. "Well, these guys had smoked their way through Vietnam and survived all this awful stuff, and their nerves were on edge. They were confined to bed, and their lives had changed. So I would have somebody stand guard, and when the chief nurse was coming, all cigarettes went out."

Gutierrez was consistently amazed at the resiliency of her patients. "The Navy had these old wooden wheelchairs," she says. "Some of the patients would come back with half-body casts or one full leg in a cast all the way to their toes. Somehow they would get out of bed and into those wheelchairs, and then they would race around. Sometimes I would get a call from security on the base and they'd say, 'Um, we have two guys in wheelchairs here racing down the street. Are you missing anybody?'"

They were still young and playful, Gutierrez explains. "I remember one of the first times I made rounds, this young man lifted up his stump and waved to me. You knew he was doing it for shock value. But they really tried to make the best of bad situations."

Although Gutierrez served as a nurse in the Navy for more than 20 years, she never made it to Vietnam. "I never felt like I did much," she says, "until I was stationed [at the VA] in Long Beach. One of our counselors had a 'homecoming' for Vietnam vets. We lined the hallway with all 70 patients and all our staff. We held little flags and welcomed them home. Then we sang *America the Beautiful*. There was not a dry eye in the place. And the VA counselor said he really appreciated the nurses and doctors in Vietnam who had helped him, but he really wanted to thank those who were Stateside who helped him transition back to the real world. I just could not keep the tears from flowing, because that was the first time I felt like I had really done something that mattered to someone—that I had been here when they came back."

— **Gail Gutierrez**
  U.S. Navy, Lieutenant, nurse
  Corpus Christi, Texas: 1968

Bernice Scott, U.S. Army Nurse Corps, aids a medical team in treating a wounded man at the 2nd Surgical Hospital in Chu Lai. *Photo courtesy of NARA.*

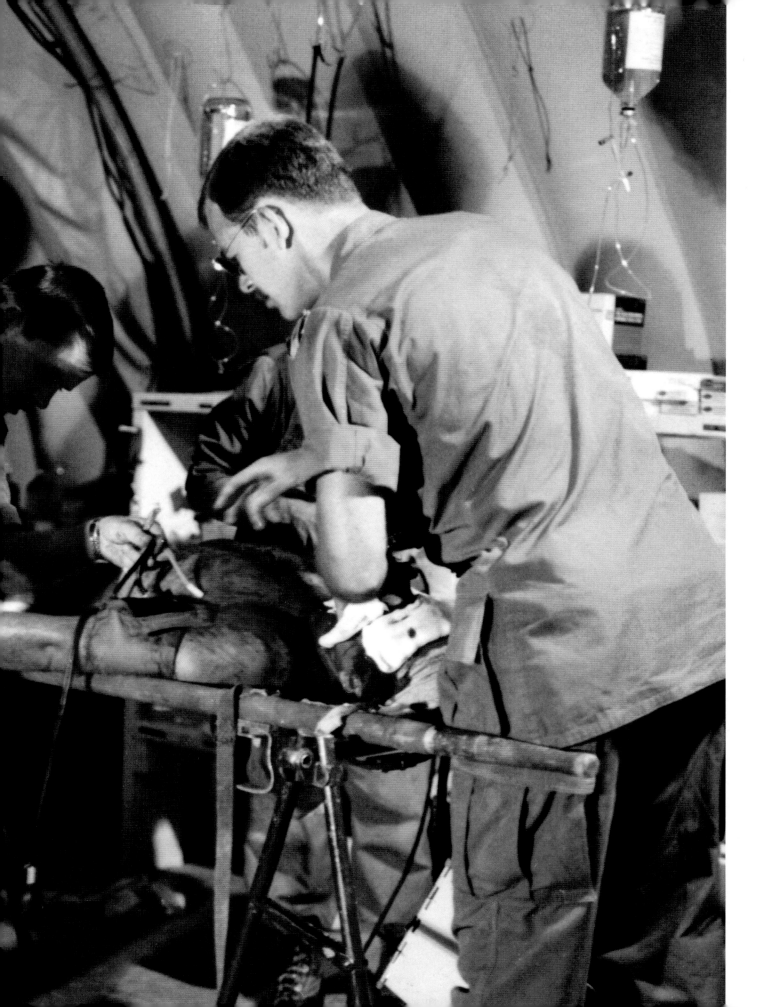

How Do You Thank Someone for Saving Your Life?

When 21-year-old Patrick Malone left Vietnam in 1968, shrapnel had shattered his spine and his hopes of walking again. But he likely would have died, if not for the unselfish courage of a fellow Soldier. It wasn't until 30 years later that he was able to find and thank him.

Malone initially wanted to fight in Vietnam to avenge a good friend who had stepped on a landmine. But that goal changed when he got to know a 12-year-old Vietnamese girl, Hanh, who did the Soldiers' laundry. He considered himself her adopted big brother.

"I no longer wanted to kill people, I wanted to understand them," Malone says. "I wanted to understand what we were fighting for, to show some compassion."

The night Malone was hurt, the base camp at Rach Kien was attacked with 375 mortar rounds, plus rockets. "And then we started receiving some small-arms fire," he says. He heard screams that the enemy was in the camp. He quickly grabbed his M-16 rifle and shot two of the attackers in the chest.

"The next thing I knew, I was hit," he recalls. "I couldn't breathe, couldn't move. I couldn't figure out what had happened. I thought I just had the wind knocked out of me. I ran my hand down my leg and felt a hole about the size of a silver dollar. I thought, I've just got a couple of holes. I'll be back."

Malone was dazed, but help quickly arrived. "Before I knew it, they picked me up and took me to the aid station," he remembers.

His friend, 18-year-old Donald Barriere, had been about 10 feet away when the mortar fragments hit Malone. "Don ran to the jeep and tried to use the mic to call for help," Malone says. "But he had a speech impediment. He stuttered, and when things got real bad, it affected him." In the chaos and panic of the attack, the speech impediment took over, and Barriere couldn't speak. "So he ran 150 yards to the aid station, through an incoming mortar attack, to get the medics and come back to me," Malone says. "It was unbelievable."

Before being flown to the hospital, Malone asked Barriere to find his "little sister" Hanh and make sure she was okay. He was still feeling foggy at Long Binh Hospital when a colonel came in and bluntly said, "Sgt. Malone, you'll never walk again." For the next 30 years, Malone tried unsuccessfully to find the friend who risked his life to get him help. In 1999, Malone and his wife, Becky, finally reconnected with Barriere. "It was a roller coaster of emotions," Malone recalls. "It's so hard to see the person who saved your life—how do you thank them for doing that? I can only imagine what it was like for Don, at 18, seeing his sergeant on the ground, and my flak jacket shredded and smoke coming off it." The two former Soldiers kept in touch until Barriere passed away in 2012.

There was another friend with whom Malone lost contact—his "little sister" Hanh had married an American Soldier and moved to Ohio. They re-established contact in 1995, and their friendship remains strong even after 50 years.

— **Patrick Malone**
U.S. Army, Sergeant, 2nd Platoon, Battery H, 29th Artillery (Searchlight),
9th Infantry Division
Vietnam: 1968

Soldiers protect their ears during the firing of an 81mm mortar. *Photo courtesy of NARA.*

A Douglas A-1E Skyraider escorts an HH-3C rescue helicopter as it flies to pick up a downed pilot in Vietnam. *Photo courtesy of NARA.*

1969
Peak Strength

President Richard Nixon took office promising "peace with honor"—a plan to bring U.S. troops home, as the South Vietnamese took over more of the war effort.

Meanwhile, the fighting raged on, with U.S. victories but also mounting casualties. The number of American combat deaths now surpassed those during the Korean War, and the Paris Peace talks remained stalled. Once their huge Easter Offensive was defeated, however, the North Vietnamese became more interested in talking.

Although U.S. military forces in South Vietnam numbered 543,000 in the summer of 1969, the reduction of forces had begun. Meanwhile, half a million anti-war demonstrators marched on Washington. The buzzwords that year were *drawdown, moratorium*, and *the silent majority*. As the year came to a close, thousands of nervous American families watched on TV to see the results of the nation's first birthday-based draft lottery. Meanwhile, in Vietnam, soldiers and grunts, sailors and fighter pilots continued to fight.

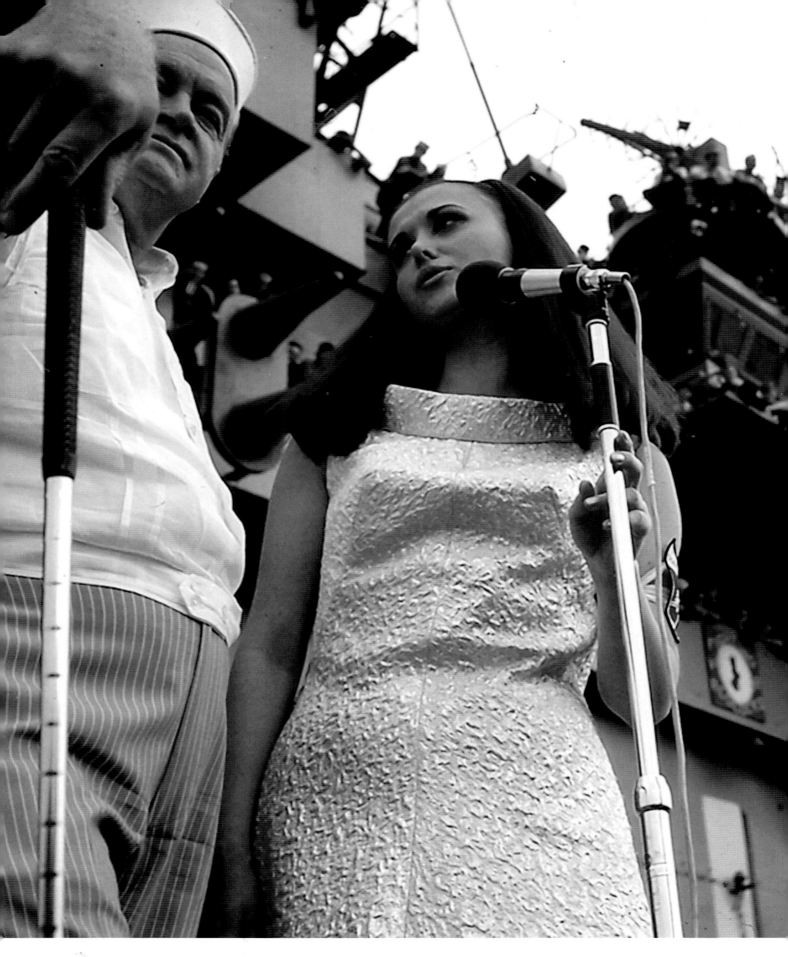

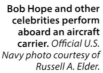
Bob Hope and other celebrities perform aboard an aircraft carrier. *Official U.S. Navy photo courtesy of Russell A. Elder.*

TIMELINE 1969

January 25, 1969
U.S.-DRV peace talks begin in Paris.

March 11, 1969
Levi Strauss starts selling bell-bottomed jeans.

March 25, 1969
John and Yoko Ono stage a "bed-in for peace" in Amsterdam.

March 28, 1969
Dwight D. Eisenhower, 34th president of the United States, dies at age 78.

April 30, 1969
U.S. troops peak at 543,482 in country.

May 10, 1969
The Battle of Hamburger Hill ends major U.S. ground combat operations.

July 20, 1969
The crew of Apollo 11 lands on the moon; astronaut Neil Armstrong takes "one small step for man, one giant leap for mankind."

August 15–17, 1969
More than 400,000 gather on Max Yasgur's Catskills farm for the Woodstock Festival.

September 2, 1969
DRV leader Ho Chi Minh dies at age 79.

September 16, 1969
President Nixon orders the withdrawal of 35,000 Soldiers from Vietnam.

October 15, 1969
Millions across the U.S. join in the largest one-day demonstration against the war.

October 29, 1969
The U.S. Supreme Court orders immediate desegregation.

December 31, 1969
U.S. casualties in Vietnam total 40,000.

A Poem for Leo

On a rainy spring day in 1990, Glenn Baker walked along the stark wall of names at the Vietnam Veterans Memorial in Washington, D.C., looking for the Soldiers he had known. He took a piece of paper out of his pocket and laid it under the name of Leo J. Schroller, 19, from Kennedy, Texas. By then, Baker had carried the burden of Schroller's death with him for 21 years.

In 1969, when he was an infantry squad leader, Baker was ordered to take the flank position in a foot patrol. Mentally and physically spent after walking "point man" in a patrol the day before, Baker balked at the sergeant's order. Schroller, a member of the squad, piped up, "I'll do it, Sarge!" The young Soldier took Baker's place, and soon after, a firefight broke out. Within seconds, Schroller was dead.

On the paper in Baker's pocket was a poem, "A Tribute to a Fallen Warrior Named Leo," that he had written as part of his treatment for PTSD. It was his therapist's idea for Baker to put on paper the guilt and anger he had carried for so long, and to visit the Wall.

My wishes are that he didn't die in vain,
Yet, still in my mind, I carry the blame,
And so for the "Fallen Warrior" who lives no more
I pay tribute for the war I abhor.

Baker was in graduate school at Howard University when he was drafted into the Army and sent to a war he didn't believe in. His mother sewed scriptures into his shirts before he left, but as soon as he arrived in South Vietnam he was given a set of jungle fatigues and ordered to turn in his civilian clothes. "I pleaded with the sergeant, telling him my mother had sewn scriptures in my clothing and they were to be my shield and buckler," Baker remembers. The sergeant let him keep one shirt.

In the months that followed, Baker was wounded twice, received a "Dear John" letter from his fiancée, and was caught in friendly fire. He learned early on, he says, "to not think about any of it." He found racial camaraderie out on missions, but a Confederate flag flown by other Soldiers awaited when he and his black buddies would return to base. Baker remembers a day when he and his unit rounded up Vietnamese civilians in a village, circled them with barbed concertina wire, put water out of reach, took off the villagers' hats, and watched the women and old men wilt in the sun.

"I gave an old lady water and got my heels locked [reprimanded]," he says. The mistreatment "didn't rise to the level of My Lai. Just another day in Vietnam."

Twenty years later, visiting the Wall, Baker realized that in that shiny black marble, "you see yourself in the stone," both literally and metaphorically. Even now, the war is still very real to him. "I want to feel proud for serving my country," he says, but he wants Americans "to remember the war honestly," warts and all. "We owe that to ourselves and to our children."

— **Glenn Baker**
  U.S. Army; Private First Class, infantry, 1st Cavalry Division (Airmobile)
  Vietnam: 1969

The 1st Cavalry Division (Airmobile) suffered more casualties in the Vietnam War than any other U.S. Army division: 5,444 men KIA and 26,592 WIA.

Sergeant Pedro Marfisi, Company D, 3rd Battalion, 8th Infantry, 4th Infantry Division, inspects the jungle canopy. *Photo courtesy of NARA.*

The Stained-Glass Window

During John Fraser's second tour as a chaplain in South Vietnam, he was stationed at the 8th Radio Research Field Station, Phu Bai, where he felt lucky to have a little chapel—even if it was in a dilapidated building and the church bell sounded like drumsticks on a dishpan. Someone had nailed plywood on the inside of the shutters to keep out the rain.

What the place needed, he decided, was a stained-glass window—but of course that was too expensive, and besides, if the Viet Cong dropped a mortar round or a rocket close by, the window would shatter. So Fraser ordered some colored acetate from the United States. His plan was to put the acetate over a sheet of plexiglass and then separate the colors with duct tape. But plexiglass was as scarce as hen's teeth, as they say in Georgia.

"So I went to the supply sergeant and said, 'What would I have to do to get enough plexiglass to put into the windows of that chapel?'" Fraser recalls. "And the sergeant said, 'Chaplain, if you'll get me two fifths of whiskey, I'll get you enough plexiglass.'"

So Fraser went to the PX Class VI store, picked up two fifths of Old Grand-Dad, and stood in line at the cashier stand. In front of him was a young Soldier who saw the whiskey and started to say, "Ahh… you're going to have a good evening!"—but then noticed it was the chaplain, and quickly stopped.

Fraser said, "Son, do you remember that story about Jesus turning the water into wine?" And the Soldier said, "Yes sir, I do!" And Fraser responded, "Well, I'm going to turn this whiskey into a stained-glass window." And he did.

"The troops loved it," he says, and attendance at the chapel increased.

— **John R. Fraser**
U.S. Army, Major, chaplain, 8th Radio Research Field Station, Signal Corps
Vietnam: 1969–70

Literally a Wingman

It was one of those socked-in days in the Marble Mountains, with clouds low and thick. On the ground some 20 miles away lay a wounded Marine who was going to die if he couldn't get airlifted to a hospital as fast as possible, but the poor visibility prevented the emergency medevac helicopter from landing.

Dan Baker was in a AH-1 Cobra helicopter gunship, providing backup for another Cobra piloted by Capt. Roger Henry and Lt. Dave Cummings, which in turn was providing weapons backup for the medevac CH-46 Sea Knight that couldn't land. Then, over the radio, Baker heard that Henry and Cummings were going to attempt to evacuate the wounded Marine themselves.

Despite just 20 feet of visibility, Henry and Cummings were able to maneuver their small helicopter onto the ground. But there was a problem—a Cobra is only three feet wide and seats just two passengers, the pilot and the gunner. The wounded Marine brought the total up to three. Baker waited in the clear air over the lowlands, wondering what was going to happen next.

Before long, though—and this was a sight that more than four decades later Baker can't forget—the Cobra emerged through the clouds, with the wounded Marine strapped into the back seat and Cummings outside the cockpit, astride one of the rocket pods attached to the chopper's stubby wing.

"It's the middle of Vietnam, in enemy territory, zero visibility," Baker remembers. "And all of a sudden I see him straddling the wing like a pony. He was facing forward, holding on for dear life, probably going 120 miles per hour. And he gives me a victory sign."

"We did so many dangerous things," says Baker, one of the first Cobra pilots in Vietnam. "Not because we were brave, but because they needed doing."

— **Dan Baker**
U.S. Marine Corps, 1st Lieutenant, helicopter pilot, 367th Marine Light Attack Helicopter Squadron
Vietnam: 1969

Early-model Cobra gunship
flies in close air support.
Photo courtesy of NARA.

ARVN troops advance on Viet Cong position as the VC attack and burn areas of Saigon during the Tet holiday. *Photo courtesy of NARA.*

Goin' Alone

When Chuck Lyford landed in South Vietnam in 1969, he was issued an M-16 rifle, a .45 semi-automatic pistol, and a stethoscope. When they called out his assignment—battalion surgeon for the 299th Engineers—he heard someone nearby whisper, "Oh, man." Lyford learned that the 299th had suffered a 50-percent casualty rate during the Tet Offensive in 1968.

As a battalion surgeon, Lyford was uncertain what he was supposed to do—most of the wounded Soldiers bypassed the unit and went straight to the hospital, and his commanding officers didn't provide much direction. So Lyford carved out his own responsibilities. His unit was stationed right outside the village of Qui Nhon, so he saw U.S. and local Vietnamese patients at the clinic there. He became the public health officer, which included inspecting local kitchens. And he supervised the medics assigned to each of the 299th Engineer companies scattered throughout the central part of South Vietnam.

Supervising the medics, of course, meant traveling to their locations. Lyford's driver explained how they would make those trips. "Doc, this may sound strange," the driver said. "I'm going to drive, and I'm not gonna keep the speed limit, and we're goin' alone. And we're gonna take a jeep, so Charlie won't bother us. He won't risk giving away his position on a jeep. If we go in a convoy, we'll get pinned down."

So that's how the pair traveled.

Morale was low in the battalion, so when a new commanding officer arrived, he asked Lyford to help out. "He said, 'See those 55-gallon drums over there? I want you to take three of them, cut them in half and put legs on them.' And every Saturday night we had barbecued steaks," Lyford recalls.

He got the steaks from the battalion cook by giving surplus penicillin to the cook, who in turn used the medicine as trading material with other units. "It was the Army form of the black market," Lyford says. "We called it 'midnight requisitioning.'"

When his unit moved to a safer location, Lyford took on yet another responsibility—stopping at local villages to offer medical help on the way to visiting other companies. "Whoever showed up, we treated," Lyford recalls, noting that many of his patients were the mountain people, or Montagnard, an indigenous, non-Vietnamese tribe. "They were very trustworthy, letting us know about Viet Cong movements."

Decades after returning home, Lyford's feelings about the war are strong. "In '69 and '70, it was very obvious that we had won the war in Vietnam," he says. "But we got defeated on the streets of California and New York."

— Charles "Chuck" Lyford
U.S. Army, Captain, battalion surgeon,
299th Engineer Battalion, 937th Engineer Group (Combat)
Vietnam: 1969

Engineman 3rd Class Robert Sands, from the Coast Guard Cutter *Point Glover*, applies first aid to the injured hand of a child on Tamassou Island in the Gulf of Thailand. *Photo courtesy of NARA.*

A Black Shaman Hat

When Navy corpsman Perry Melvin arrived in South Vietnam, he was assigned to the Marine Combined Action Program (CAP), whose squads lived and worked alongside the Vietnamese people in their own villages. Melvin became so respected that the local villagers gave him a black hat worn only by the shamans. They called him *bocsea*—Vietnamese for doctor.

When Melvin wasn't on patrol with the unit, he spent much of his time working with the local midwives and shamans. "Before heading over to Vietnam, I spent a little over a year in a Newport [Rhode Island] hospital, in the nursery and maternity and pediatrics," he says. "I wondered what I was going to do with that. But I used it in Vietnam. I had the opportunity to work with a lot of people and build their trust."

During the day, the CAP unit went on patrol, set up ambushes, and worked with the local people to increase protection and decrease their terror. "We had two or three villages that we moved between, trying to keep the enemy guessing where we were," Melvin says. "The enemy hunted us at night; we hunted them during the day. The Viet Cong were stealing [the villagers'] food, terrorizing them, in some cases even executing them. These people were innocent, caught between us and the Viet Cong."

Because CAP teams were smaller—only eight to ten men—they were very effective at finding and killing the Viet Cong and helping the local people. "We lived with them, we smelled like them, we ate their food, and we learned to move like them, quietly and quickly," Melvin explains. "Larger units got hit constantly—we were much less likely to get hit."

But it did happen. When Melvin received his CAP assignment, he was replacing a corpsman on a unit where all but one of the members had been killed. "It was profound," he recalls, "but the sergeant who took over the unit was experienced, and we got our wits together and we won the hearts and minds of the people."

Melvin himself was also injured only a few months into the assignment. He was medevaced out—leaving his few possessions, including the treasured black shaman hat, behind.

— **Perry Melvin**
U.S. Navy, Petty Officer 3rd Class, corpsman, Marine Combined Action Program (CAP)
Vietnam: 1969

U.S. Navy corpsman eats chow from a can while he serves in a landing zone near Dong Ha. *Photo courtesy of NARA.*

The Best and Worst Day

George Darrough's worst day in Vietnam was not during one of the 50 missions where his A-37A Dragonfly light attack aircraft was shot at, or even the day he flew his plane through a fireball from a napalm blast. His worst day was when he heard about the teacher at Phan Chu Trinh School.

Darrough was a fighter pilot, as gung-ho as all the other fighter pilots he knew. They were all Type-A personalities, he says, but not like the cocky pilots you see in the movies—not at all like Tom Cruise in *Top Gun*. They were just highly motivated, he explains, especially if they were providing air support for South Vietnamese troops on the ground who were in trouble.

His squadron also helped build a school in Bien Hoa, after the town was shot to pieces during the Tet Offensive in 1968. It was a one-room school, maybe 25 by 40 feet, and on the day the school opened, about a hundred kids showed up. There was one teacher, a "kind, wonderful Vietnamese woman," he says.

On the second day of school, Darrough got a devastating call. The Viet Cong had killed the teacher and nailed her body to the wall of the school—to send a warning message to the villagers.

But that tragic day was also Darrough's best day, he says. "The thing that amazed me so much was that the villagers took her body down and cleaned her up, and her husband said, 'School will start again tomorrow morning at 8:00.' His love for his wife, the schoolchildren, and his country exceeded his fear of the same fate. "He wanted to send a message back to the Viet Cong. Certainly, none of us would have had that much courage."

The husband took his wife's body home and buried her, and came back to teach the class. "And he lasted," Darrough says. "The school was still operating when I returned home several months later. It was that kind of commitment that gave us the reason to keep on fighting the Viet Cong."

— **George N. Darrough**
U.S. Air Force, Captain, fighter pilot
Vietnam: 1969–70

The Saddest Mail-Call Day

The letter arrived, as many did that year, smelling like perfume. The envelope was aquamarine and the handwriting was perfect, and when Dean Glorso saw it in the pile of mail, he felt awful.

Glorso was the mail clerk for a squadron of Marines at the Da Nang Air Base. He felt lucky to be away from combat, plus "I got respect from everybody," he says, "because no one wanted their mail messed with. Mail was their lifeline back to the States."

But the job came with its own sorrow. On Glorso's saddest mail-call day, September 30, 1969, he knew that the familiar letter in the flower-scented blue-green envelope was for the squadron's executive officer. The wife of the second in command had written her husband before she learned that he was soon to be declared missing in action.

Later, Glorso wrote a poem about the incident, "The Major's Mail Call," reflecting on how happy Major Luther Lono always was to get letters from his wife, and wondering how many years she would weep "…not knowing the fate of her man so dear. Missing in Action must be every love's fear." Lono's remains were finally found in 2000, where his A-6 Intruder went down during a night armed reconnaissance mission in Laos.

— **Dean Glorso**
U.S. Marine Corps, Corporal, mail clerk and administration, 1st Marine Aircraft Wing
Vietnam: 1968–69

Both Sides of the Story

William Currlin remembers trying to sleep on the ground with his troops during South Vietnam's monsoon season. "It lasted about five months," he remembers. "There was rain or drizzle, something every day. You'd lie in the mud and put your poncho over your face so maybe you could sleep a little bit. It was just a miserable feeling."

Currlin, an Army lieutenant, was in Quang Tri Province when something worse than rain awakened him one morning. His unit's radio operator, horrified, began slapping at Currlin's back. "I was covered with little baby scorpions, maybe a quarter-inch long," he explains. "I'd been sleeping with them all night. We got going because we figured their mom must be close. There were snakes and insects. A bum in the street sleeps better than that."

Because the men had to carry water with them, the priority was on drinking water, not water for hygiene or grooming. "When I first took over my platoon, the guys saw me sniffing at something, and it was them," Currlin says. "It wasn't long before I smelled like that, too. Even brushing teeth took precious water. Shaving was hard to do."

Vines caught the men's arms, faces, and clothing as they walked. "We got scraped a lot, and because we were dirty, our wounds got infected really fast," Currlin recalls. "Some guys got cellulitis, where the infection goes down to the bone. Sanitary conditions were terrible. But in spite of all this, we would go on."

Currlin and his men never let their discomfort get in the way of their mission to search and destroy. "This one time when we were walking patrol, we were attacked," he says. "We walked into a bunker complex dug into a mountain. They were firing at us. We had to take cover, so we were behind a hill, taking fire pretty bad. We had used up our allotment of artillery." His commanding officer told Currlin to attack the bunker complex. "It was a suicide mission," he recalls thinking. "We had no cover. But we obeyed the order because that's what you did. If you are told to do something, you do it. So we decided we would come in from the side."

After about 25 minutes of firing on and returning fire from the bunker complex, the battleship USS *New Jersey* (BB-62) was able to shoot 16-inch diameter shells at the location. The enemy kept firing until two U.S. Air Force jets dropped bombs on the bunker area. "We followed blood trails, looking for enemy wounded, for a week," Currlin says. The action earned him the Bronze Star Medal for valor.

In the last few months before Currlin's return to his native Connecticut, the Army assigned him to work in South Vietnamese villages. It was the first time he had direct contact with any Vietnamese other than the enemy, and it was an eye-opening experience for him. "It was good for me because I had developed a hate for anyone with Asian eyes," Currlin says. "They [the enemy] had killed and tortured our people. But I got a chance to see civilians and children, and got to see humanity very similar to what we have. It was a good way to complete my tour. That way, I didn't walk away with the hate others did."

—William E. Currlin
U.S. Army, 1st Lieutenant, platoon leader, 1st Battalion,
501st Infantry (Airborne), 101st Airborne Division
Vietnam: 1969–70

Troops of 2nd Battalion, 12th Infantry, 25th Infantry Division move through the rising mist as the second day of Operation Bushmaster begins. *Photo courtesy of NARA.*

Gun firing on USS *New Jersey* (BB-62). *Official U.S. Navy photo courtesy of Russell A. Elder.*

A Salute and a Pow Wow

After Robert Poolaw was wounded in the chest and hospitalized in Da Nang, South Vietnam, he was eager to get back to his job as company commander. But he was warned that he'd first have to convince his battalion commander that he was in good enough shape to lead his men—which Poolaw knew would be a hard sell because he could hardly even raise his arm to salute. As a Native American, Poolaw was proud to follow the military tradition of all the men in his family who had fought on battlefields in World War II and Korea. The last thing Poolaw wanted was to be reassigned to a desk job.

So he came up with a plan: Poolaw and his gunny,* who had also been hospitalized, would practice saluting (in case their first encounter with the battalion commander was outside) and shaking hands (in case the first encounter was inside), and they'd do both tasks 300 times a day for the three days until their release from the hospital. When that proved difficult, Poolaw came up with Plan B: 50 salutes and 50 handshakes a day. When the gunny was released early, Poolaw went to Plan C: he saluted himself in the mirror. And he shook hands with himself, over and over and over. By the time he left, his salute and his handshake were almost perfect— good enough to get his company back.

Poolaw served two tours in South Vietnam, commanding two companies. When he returned home he wasn't jeered or ignored, as were so many Vietnam vets. Like many other Native Americans who served in Vietnam, he was welcomed home at an all-day pow wow, with dances and singing in his honor. People from seven tribes showed up. They gave him so many gifts he had to give most of them away.

The elaborate welcome-home ritual was their way of saying "we're glad you went and we're glad you came back."

— **Robert Poolaw**
U.S. Marine Corps, Captain, Special Landing Force
Vietnam: 1966–67, 69–70

Marine slang for a Gunnery Sergeant (E-7)

Goooood Morning, Vietnam!

It was a phrase I shouted virtually every weekday at 6 a.m. from the studios of the American Forces Vietnam Network in Saigon between October 1968 and December 1969. I wasn't the first to use those memorable words—that was Adrian Cronauer, who was famously portrayed by Robin Williams—but that became the signature sign-on of every early-morning DJ on AFVN.

While I am proud of my service, there is one small incident that fills me with more than a little bit of embarrassment. President Richard Nixon had taken office in January [1969] and was preparing to make his first holiday address to the nation. …In those days, there was no technology to allow for live television coverage to Southeast Asia, so the address was delivered to us by radio. Because of the time difference, Nixon's prime-time address was to take place during my morning show.

The process was a simple one…. When it came time for the president to start his speech, I would hear him being introduced through headsets. Right on schedule, the CBS announcer began his introduction and I broke into the music I was playing to announce in the most important tone I could muster, "We now go to Washington for an address by President Nixon. Ladies and gentlemen, the President of the United States."

I don't remember much about what he said, but it was effective and occasionally moving, and by the time he was wrapping up, I felt he had done a very nice job. When he came to the end and began shuffling the papers in front of him, I flipped the switch in the other direction, and again in my best announcer's voice, I told everyone they had been listening to the Commander in Chief, and then it was back to the business of entertainment.

After I started the next record, I wanted to hear what the CBS announcers back home were saying about the speech…. I flipped back to the CBS feed in the studio where, to my horror, I discovered they weren't discussing the speech because the president was still speaking! Apparently, what I deemed to be an effective close was merely an effective pause…. To make matters worse, I heard Nixon say, "And now I'd like to speak directly to the men and women serving our country in Vietnam."

I had a quick decision to make. Should I jump back on the air and confess that I had cut off the leader of the free world in the middle of his address, or should I just keep playing music and hope for the best? It was as if a little angel was perched on one shoulder with a little devil on the other. The angel, of course, was right. The president was speaking and it was my duty to reconnect him.

But I had to admit that the devil was making some good points. His main argument went this way: Because the CBS feed was coming directly into the AFVN studio, and I was the only one monitoring it, I was literally the only human being in the world who realized that the people the president thought he was speaking to couldn't actually hear him. So, really, what was the harm?

True, he was sending holiday greetings to the troops and promising to bring them home soon, but they were already listening to the 1910 Fruitgum Company singing 1, 2, 3 Red Light. Heck, now I'd be cutting off that fine song in the middle, and two wrongs don't make a right, do they? And by the time I explained what had happened, he might be finished anyway. In short, the devil made me not do it.

So, it is with pain and embarrassment that I confess that my comrades in Vietnam never heard the president's words to them back in 1969. So, very belatedly, I want you all to know that Richard M. Nixon wishes you a very merry Christmas.

There. I feel better.

— Pat Sajak
U.S. Army, Specialist 5, DJ, American Forces Vietnam Network
Vietnam: 1968–69

Sajak, Pat. "Goooood Morning Vietnam!" *On Patrol* magazine, Summer 2014.

U.S. Army UH-1H Huey helicopters, with an AH-1G Cobra trailing behind, return from refueling at Bu Dop Special Forces camp. *Photo courtesy of NARA.*

Soldiers attach a sling-load of fuel drums to a CH-54 "Skycrane" helicopter during a recovery operation. *Photo courtesy of NARA.*

Helicopter crew chief loads 20mm cannon shells aboard his Cobra helicopter. *Photo courtesy of NARA.*

That Night Changed My Life Forever

During the battle of Hamburger Hill in May 1969, casualties seemed to flood the hospital at Tachikawa Air Force Base in Japan where Air Force nurse Linda Schwartz was stationed. "There were so many casualties coming up from Vietnam that we were getting casualties who still had their field dressings on," she recalls.

Typically, Schwartz notes, wounded Soldiers received initial follow-up care in South Vietnam before arriving in Japan, where patients stayed until they were stable enough to make the long journey back to the United States. But one night at the height of fighting on Hamburger Hill, the field hospitals were overwhelmed. "They were just reinforcing the bandages rather than replacing them because there were so many wounded," she says. "And then that night we had to clear out the hospital of people who could be moved because we needed the beds."

When the first casualties from that battle started coming, Schwartz was stunned by what she saw. "They had chest wounds and they had chest tubes in, but they were walking because they were not the worst casualties," she says, recalling her astonishment. "Their buddies were helping them. And that experience of people with such really bad gunshot wounds and injuries walking and their buddies helping me, it really changed my life forever. Seeing this great spirit and this taking care of each other as they had, I knew that I would never be able to go back to civilian nursing where little old ladies need their aspirin and they can't wait five minutes. That is when I really decided that I wanted to make the military my career."

Schwartz was already a registered nurse when she decided to enlist in the Air Force. "The war in 1967 and 1968 was really heating up," she explains. "And at that time I just felt that I was so blessed to be here in the United States, that I had a skill that could be used, and I joined."

— **Linda Schwartz**
U.S. Air Force, Captain, nurse, 6407th Hospital
Japan: 1968–69

In September 2014, Dr. Linda Spoonster Schwartz was appointed Assistant Secretary of Veterans Affairs for Policy and Planning.

A Day of Loss in Tay Ninh

On a morning in 1969, medic Richard DeLeon rode in a helicopter with a load of supplies, recruits, and a priest. DeLeon was returning to active duty after being hit by shrapnel in a Viet Cong ambush just a few weeks earlier.

The helicopter flew DeLeon to Company A's location near a North Vietnamese bunker complex in Tay Ninh Province. The location was surrounded by a dense bamboo forest and had been taken in a fierce battle. "This bamboo is so hard you can't cut it with a knife, you can't break it with a machete," DeLeon explains.

By chopping at stalks and burning the stumps, the men had cleared enough space for a helicopter to hover but not land. DeLeon safely dropped the eight or so feet to the ground, and he helped the priest disembark before heading off to greet his men.

"They introduced me to my new lieutenant, a very nice man who wore a baseball cap," DeLeon remembers. "'Doc, glad to meet you,' he said. And I said, 'How are my boys?' And he said, 'I think we are all going to be fine.'"

Just before DeLeon landed, a crate of grenades had dropped from another helicopter and split open. The men had scrambled frantically to gather the 30 or so grenades before DeLeon's helicopter arrived. "Fires were still smoldering from where they had tried to burn the bamboo," explains DeLeon.

The men had missed one grenade from the earlier drop. After greeting DeLeon, the lieutenant headed back to the bamboo clearing, reaching it just as the heat from the smoldering fires detonated that lone grenade. "Suddenly, the men called out, 'Doc, your LT has been hit!'" DeLeon says. "I went and found him. I had just met him. He had a big hole in his head—he'd died instantly, probably mercifully. So now the helicopter is coming back in, and I put two people up into it: the padre—who was now trembling and shaking; I couldn't even console him because he had seen what happened to the LT—and we wrapped up the LT and put him on."

As another helicopter approached the clearing on its last run of the day, DeLeon heard 30-caliber machine-gun fire. The chopper sputtered, then crashed. On board were seven replacements, two gunners, and two pilots. DeLeon's platoon rushed to the downed ship, returning the Viet Cong fire. Only two men on the helicopter made it out alive; both were severely wounded.

DeLeon remembers looking in disbelief at the retrieved bodies, wrapped in poncho liners,* awaiting transport out. "These boys never even saw a lick of combat," he says. "The next thing that I recall is sitting around with my platoon, nobody saying anything, waiting for the order to saddle up and head out. We were going on another mission."

— **Richard R. DeLeon**
U.S. Army, Specialist 4, medic, Company A, 1st Battalion, 12th Cavalry,
1st Cavalry Division (Airmobile)
Vietnam: 1969

Rubberized cloth ponchos were issued with a soft, quilted liner for extra warmth and softness.

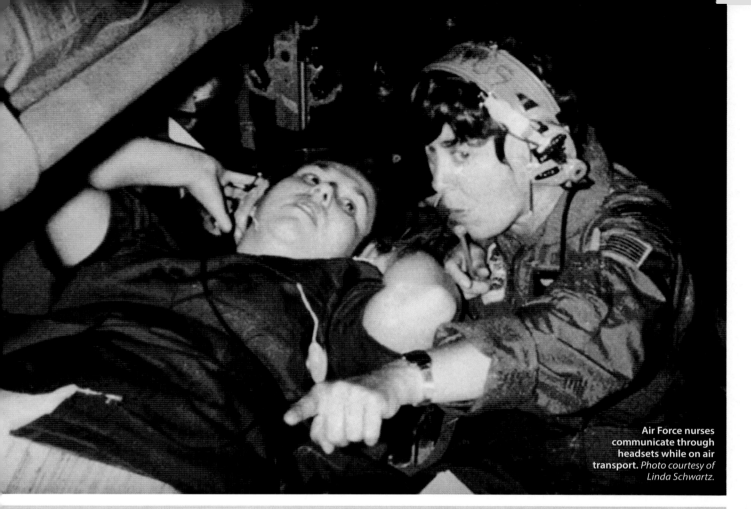

Air Force nurses communicate through headsets while on air transport. *Photo courtesy of Linda Schwartz.*

The Gold Watch

It was Marlene Bayer's first week in the 12th Evacuation Hospital at Cu Chi, South Vietnam, and the staff was looking forward to a milestone. By February 1969, the emergency room had treated almost 10,000 patients.

"The personnel were planning ahead for the arrival of the 10,000th patient," the nurse recalls. "They were going to give that patient a gold watch to commemorate the event." But when the 10,000th patient arrived in the emergency room, both his arms had been blown off in combat.

"The staff was so emotionally affected they quietly put the watch away, and I don't know whatever became of it," Bayer says. "To my knowledge, they never again tried to commemorate a certain number of patients." But she quietly kept track, and over the course of her year in the wards, the emergency room had treated about 20,000 more patients.

Bayer remembers hoping that as time went on, they would have fewer wounded Soldiers to care for. Instead, the steady stream of patients kept increasing. "At the time, it seemed as if nothing we did made any difference," she says. "But I've been told by the veterans we cared for that we did make a difference for them."

Many years later, Bayer went to view the traveling Vietnam Women's Memorial as it passed through her hometown of Wichita, Kansas. The veteran overseeing the sign-in sheet thanked Bayer for her service and mentioned that he was injured near Cu Chi the day she arrived in country. The nurse and the Soldier shared a sweet moment when they discovered they had both been in the 12th Evacuation Hospital in February 1969.

"Those G.I.s were wonderful patients," Bayer says. "Almost always their first concern was for their buddies, not themselves. Despite their severe injuries, they asked for very little pain medication. They were tough, resilient, and grateful for our help. Caring for them was an honor and a privilege, and the most satisfying time of my nursing career."

— Marlene K. Bayer
U.S. Army, Captain, staff nurse, 12th Evacuation Hospital
Vietnam: 1969–70

An infantryman is lowered by member of the recon platoon.
Photo coutesy NARA.

50 Feet Above Ground

By the time Tony McPeak got to South Vietnam in 1969, he was used to flying low and fast—skills that came in handy in top-secret missions along the Ho Chi Minh Trail in Laos.

When he arrived in Vietnam, McPeak flew combat missions in an F-100 fighter with the 37th Tactical Fighter Wing (TFW) at Phu Cat Air Base. But with his unique previous experience as lead solo pilot in the elite U.S. Air Force Thunderbird aerial demonstration team, he was soon assigned to Detachment 1, 416th Tactical Fighter Squadron (TFS), a specialized forward air control unit with the call sign "Misty" (named by the squadron's first commander after his wife's favorite song).

The detachment's mission was to locate enemy vehicles and supplies—and they did that by conducting high-speed, low-level reconnaissance flights as low as 50 feet at 500 knots over the Ho Chi Minh Trail. "You can't find anything if you're up at, say, 1,000 feet or so," McPeak explains. "If you see something that looks suspicious, you've got to go right down on the road to look back under the trees" for evidence. For example, he explains, perhaps the enemy had turned off the road to hide a truck during the daytime, in preparation for moving supplies during the night.

"What you looked for was camouflage netting on the top of some trees, or some kind of regular lattice work. The jungle itself is never regular, so if you see anything that's geometric, you know that's man-made," he says. The "Misty" pilot would circle back down, going low enough to get a good look, and then call in fighter jets to bomb the area.

McPeak became the detachment commander in April 1969. After 98 "Misty" sorties, he went back to flying routine attack missions with the 416th TFS, then moved to a position on the 31st TFW staff. By the end of his tour in Vietnam, he had flown a total of 269 combat missions.

Because it was considered more dangerous than other kinds of combat sorties, he says, the "Misty" FACs were limited to a four-month tour. But McPeak was not new to danger: the peacetime Thunderbirds lost a third of their solo pilots. By comparison, he says, "combat was relatively easy."

— **Merrill Anthony "Tony" McPeak**
U.S. Air Force, Major, pilot, Commander, Detachment 1, 416th Tactical Fighter Squadron, 37th Tactical Fighter Wing; 31st Tactical Fighter Wing
Vietnam: 1969–70

General Tony McPeak served as Chief of Staff of the U.S. Air Force (1990–94) during Operation Desert Storm.

Through the Eyes of the Vietnamese People

When Rick Kiernan left South Vietnam after serving on the ground as an infantry advisor to the South Vietnamese Army, the Vietnamese people he'd worked with side-by-side every day for a year gave him a gift.

"They'd taken an artillery shell and hammered it out into a vase," Kiernan says. "And it's engraved, 'To my best friend, Captain Kiernan.' So I have a big, heavy, brass piece of the war that could have been bellicose, and they turned it into something beautiful."

Kiernan's war memento may be as unusual as his war experience. An ROTC graduate from the Virginia Military Institute, Kiernan served as one member of a five-man team that ate, drank, slept, and lived right in a Vietnamese village. Kiernan was hand-selected for his assignment, in part because he'd had six years of French in high school and college. Before being sent to Vietnam he trained in advisory techniques and operations at Fort Bragg, where he studied about Vietnam and learned Vietnamese to complement his French.

"The vast majority of Soldiers served in U.S. military units, living all together in an Americanized community, kind of like moving San Diego to Saigon," Kiernan says. "Not us. We ate chicken and rice every day, their water hole was our water supply, we lived in a hooch, and when we went to bed at night, we knew that our lives were in their hands. We lived their lives. I saw the war through the eyes of the Vietnamese people."

Kiernan spent days with his Vietnamese counterpart walking combat patrols, logging helicopter time, digging foxholes, and gathering intelligence about the North Vietnamese and Viet Cong to share with top military officials over the area. In the evenings, he returned to the village, where he taught the children English, helped the villagers with basic survival skills, and simply spent time with the people he'd grown to care for. "Most Soldiers go out every day looking for an enemy," Kiernan says. "I went out every day and found friends—victims of a war they didn't understand."

— **Rick Kiernan**
U.S. Army, Captain, infantry, advisor to South Vietnamese Army, Military Assistance Command, Vietnam
Vietnam: 1969–1970

Colonel Kiernan served as a Pentagon spokesman during the first Gulf War. After retirement, he was director of press operations and public information for the 1996 Summer Olympic Games in Atlanta.

The Sniper

One week during the fall of 1969, Steve Suttles sat in the same spot for five days, waiting to pull the trigger of his rifle. He and his spotter were hiding about 400 yards above a South Vietnamese village, where military intelligence had indicated that a North Vietnamese Army colonel would be meeting with local Viet Cong officers. Just in case he might be spotted, Suttles wore a conical hat and loose black clothes, and made sure to eat rice so that even his body odor wouldn't reveal him as American.

Suttles was the shooter that day, assigned to kill the colonel's bodyguards so the colonel himself could be captured. Suttles' spotter was a South Vietnamese Marine. Spotting is actually the more difficult and important job, Suttles says, because the spotter has to do the calculus of wind speed and distance—the shooter just has to have a steady hand. By then, Suttles was already training South Vietnamese Marines as snipers, even though he'd only been in country for six months.

As they waited, they rehearsed their plan. "We went over and over and over that routine, and I kept dry firing," quietly pulling the trigger of his empty rifle, probably 5,000 times in all. "You have to rehearse it, so that when you do it for real your mind and body fall into sync," he explains. When the NVA colonel finally showed up, Suttles quickly killed two bodyguards, and South Vietnamese Marines dressed as Viet Cong soldiers swooped in and captured the colonel and the Viet Cong officers.

When Suttles brought their captives back to the An Hoa Combat Base, he remembers, "we must have presented quite a sight, based on the incredulous looks we got. One American in black pajamas and cone hat carrying a sniper rifle, with one uniformed NVA and four VC gagged and bound with six pajama-clad escorts."

Suttles spent his 13 months in Vietnam in An Hoa, but occasionally was sent into Laos and Cambodia, although official word at the time was that no ground troops were in either country.

The longest successful shot he ever made was 1,250 feet, from a watchtower on Hill 55. "Even if I had missed, I would have done my job," he says. "I took shots knowing it would be a possible miss, just to get them looking over their shoulders, so to speak. You take them out of their comfort zone. They slow down, show themselves less, change travel routes and habits, all of which help change the dynamic of the combat zone. It lets them know somebody else is playing in their backyard."

— **Stephen "Steve" Suttles**
U.S. Marine Corps, Corporal, sniper, 1st Battalion, 5th Marines, 1st Marine Division, attached to South Vietnamese Marines
Vietnam: 1969–70

A 7th Marines sniper focuses in on a distant enemy with his Remington 700 sniper rifle, while his spotter watches through binoculars. *Photo courtesy of NARA.*

Finding Courage

Bill McClung arrived in country in the summer of 1969 and was assigned to a recon platoon. "Excuse me, sir, I think you made a mistake," he told the sergeant. "This is an infantry division." McClung, a medic, had pictured himself working in a hospital in Vietnam, in a 9-to-5 job, sleeping on clean sheets. The Army had other ideas.

There were, instead, 30-day stretches in triple-canopy jungle, trudging through monsoons, sleeping in the mud, doing battle with leeches. There was malaria, hepatitis, and jungle rot. Every evening he would scrape the skin off his Soldiers' feet, hoping to ward off infections.

His worst situations in the Vietnam War, he says, were those times when he was the only medic and there were multiple, simultaneous casualties, each needing his immediate help. "Because what you have to do, and I don't want to be melodramatic, is play God. You have to triage them, deciding which man to treat first and which man is too close to death to save."

A memory: Three men have been wounded in a firefight. The first has a small wound in his upper chest, with only a trickle of blood, and he assures McClung that he's fine. The second Soldier has a stomach wound; his guts are hanging out, his skin is chalky, and he's slipping in and out of consciousness. The third Soldier has a bone sticking out of his thigh and is screaming in pain. Who does the medic treat first? The answer: McClung turned over the Soldier with the small chest wound and discovered it was what they called a "through and through"—he would quickly bleed out and die without immediate attention. As cruel as it was, says McClung, he had to choose saving this Soldier. The man with the stomach wound was less likely to survive even with immediate aid. "It was a terrible decision that had to be made, and it haunts me to this day."

When he got home from the war, his nightmares would wake him up screaming and punching. The morning after his wedding night, he discovered that his wife had retreated to the couch. "I got out of there before you killed me," she told him. The nightmares subsided eventually, but like a lot of veterans, when he goes to a restaurant he'll always pick a seat where he can have his back to a wall.

At first he wondered if he was brave enough to treat Soldiers in combat. But today, when he gives tours at the New Jersey Vietnam Veterans Memorial, he tells students that his year in Vietnam was the most valuable experience of his life. "You do things you never thought you could do," he says. "You find courage you didn't think you had."

— Bill McClung
U.S. Army, Specialist 5, medic, reconnaissance company, 1st Cavalry Division (Airmobile)
Vietnam: 1969–70

Ambulances line up at Tan Son Nhut Air Base to evacuate patients via aircraft to U.S. hospitals.
Photo courtesy of NARA.

A U.S. Marine corporal with heavy shrapnel wounds in his legs, is given treatment in the shock and resuscitation ward of the amphibious assault ship USS *Tripoli*. *Photo courtesy of NARA.*

U.S. Air Force security police disperse along the perimeter of Tan Son Nhut Air Base in response to a Viet Cong attack. *Photo courtesy of NARA.*

What I Went There to Do

As section chief and head combat medic in an Army unit assigned to work among native South Vietnamese, John Swan performed medical procedures far beyond the basic training he'd received. "Every day was a challenge," says Swan, who learned Vietnamese and several other dialects. "But I knew that even though I may not know what I was doing, I was the only one who had the chance to do it."

Working out of a 100-pound medical chest, Swan incised a huge tumor on the side of a baby's head, gave daily shots to a toddler who had been bitten by a rabid dog, and treated a young teenage girl who had been exposed to napalm, the highly flammable fuel used during the war that stuck to skin and caused horrific burns. "Working with a sterile field was totally and absolutely impossible," Swan explains. "The only thing I could do for the girl was throw down a poncho liner. I had to cut away dead skin and treat her skin grafts."

He was sometimes overwhelmed. "There was some serious medical stuff I did that was way beyond my capacity," Swan explains. "It was scary in one way, but most gratifying in the other because I got to do what I went there to do, which was save lives."

Swan's decision to sign up for the Army was a long time coming. He'd grown up in a military family and was told as a boy that every Swan had served in the military, and he needed to serve too. However, by the early '60s he'd gotten involved in the civil rights movement and then become a war protestor. He attended school and was a full-time research associate at the University of North Carolina in the Psychology Department when he changed his mind and volunteered for the draft.

"In 1968, I realized I had not only the capacity but the will to go into the Army," recalls Swan, who actually talked to a therapist about reconciling his desire to not kill with the understanding that he'd probably have to kill in order to save lives. He also knew that, with his background in medicine and psychology, he'd likely be a combat medic—and combat medics had the highest casualty rating of any job in the Army.

Before Swan showed up for basic training, he shaved his head and was running nine miles a day. Within a week or two, he was an acting platoon sergeant running a 30-man outfit. At 24, he was the oldest recruit, and the younger guys started calling him Papa.

As Swan looks back at his time in Vietnam, he's philosophical. "It wasn't exactly as clean as I would have liked," he observes, "but war isn't clean. People in war aren't clean, and the insanity of war has to touch you. I've been touched by saving life and the insanity of taking life—both leave their imprint."

— **John Swan**
U.S. Army, Specialist 5, medic, Headquarters Company 2nd Brigade, 4th Infantry Division
Vietnam: 1969–71

The Boy

Chuck Creel was on duty at the dispensary at Phan Rang Air Base in South Vietnam one night when a late call came in about a casualty on the base perimeter. When he got out there and shined his lights through the barbed wire, he could see a child lying in the middle of the defensive minefield.

Creel had joined the Air Force hoping to be a photographer. Instead, he got his last choice — medic. But once he arrived in South Vietnam, he realized he liked helping people. He worked in the local orphanages and surrounding villages. This particular night, he gave aid to a boy who was probably assisting the Viet Cong by probing the base defensives for a weakness.

"We did not provide aid to whoever came through there," he says, "but because of my desire to help, I left the compound." When he reached the boy, he could see that an explosion had severed the boy's leg at the knee, and he was hemorrhaging.

"I remember how tight this little child held to me," Creel recalls. "He grabbed me around the neck and squeezed me really tight. I could feel the blood from the severed leg running through my clothes, and I knew that if I didn't immediately get him help, this child was going to die."

Creel applied a tourniquet on the leg and carefully retraced his steps — through the minefield and the surrounding woods — carrying the boy. At the base dispensary, the medics administered intravenous fluids, then evacuated the boy to a hospital.

It seemed like a relatively happy ending to Creel's heroic act, but life in war-torn Vietnam was seldom that simple. Two days later, Creel learned, the boy died of heroin withdrawal. An addict at age 10.

— **Charles "Chuck" Creel Jr.**
   U.S. Air Force, Staff Sergeant, medic, 315th USAF Dispensary
   Vietnam: 1971

Sergeant Creel later received the Airman's Medal for Heroism.

Like thousands of other American military personnel during the war, this U.S. Navy advisor shares a meal with his South Vietnamese comrades-in-arms. *Photo courtesy of DoD.*

A New American

Ed Canright was flying his UH-1 Huey for the 61st Assault Helicopter Company on a run-of-the mill "ash and trash" logistical mission in the An Lao valley—delivering C-rations, mail, and passengers—when the call came in for an emergency medevac nearby. Canright was the new guy on board the lift ship that day, the "Peter Pilot" just learning the ropes on his very first mission.

When Canright landed a few minutes later, a Soldier from Company B, 173rd Airborne Brigade was rushed aboard, and the Huey took off. Canright immediately smelled something strange coming from the back of the chopper. When he turned around, he saw that both the Soldier's legs had been blown off. "And I realized the smell I had never smelled before was the copious amount of blood," he remembers.

Canright assumed the Soldier could not survive with such horrific injuries. But sometimes there were good surprises in war: nine months later, while reading a military newspaper, Canright saw a picture of a man described as having served with Company B, 2nd Battalion, 503rd Infantry Regiment, 173rd Airborne Brigade. The man was from Italy. He'd been studying in the U.S. when he enlisted in the Army and had lost both his legs in a booby-trap explosion nine months earlier in the An Lao valley. In the picture, the man Canright helped save was standing on two artificial legs—and he was being sworn in as an American citizen.

— Edward Canright
U.S. Army, 1st Lieutenant, helicopter pilot, 61st Assault Helicopter Company, 268th Aviation Battalion (Combat)
Vietnam: 1969–70

Parallel Walls

Joan Furey's first patient in Vietnam was a Soldier with a spinal injury, a neck fracture, and bilateral chest wounds. She was terrified and overwhelmed, but she had to get over it—there were only five nurses and corpsmen taking care of 30 critical patients. The job meant shifts of 12 hours or more, six days a week, treating severe wounds, burns, disfigurement, amputations, and other life-altering injuries. The 71st Evacuation Hospital in Pleiku also treated injured civilians from nearby villages, including children and infants.

When enemy mortars and rockets hit the hospital, Furey would put on a flak jacket and a helmet and crawl from bed to bed to care for her patients, who couldn't be moved or left unattended. She was 22 years old, and most Soldiers in those hospital beds were younger than she was. Like the rest of the medical staff, she channeled her despair into doing the very best for her patients. "There was a purity of commitment to the work in Vietnam that you just never experienced anywhere else," she says.

Back home, adjusting to civilian life wasn't easy. Furey felt she had lost a part of her humanity. But in 1993, she reunited with colleagues and patients at the dedication of the Vietnam Women's Memorial on the Mall in the nation's capital. As the nurses stood on the Mall, a veteran in fatigues approached them with a big bouquet of roses. "A nurse saved my life in Vietnam and I don't remember her name," he said. "I'm going to give these to you, because I know there's someone out there you did the same for, and I want to thank you for all of us."

Years after the war, Furey took a trip back to the Democratic Republic of Vietnam with other nurses and found a country at peace, with kids playing and practicing their English. She no longer associated Vietnam with combat, casualties, and devastation. In Da Nang, she and her friends found a war memorial with the names of casualties from nearby villages engraved on a large stone. She took out a picture of the Vietnam Veterans Memorial in Washington, D.C., she had brought with her and showed it to the Vietnamese caretaker. "Ours has names on it too," she said. Somehow, those parallel walls helped Furey find the closure she needed.

— Joan Furey
U.S. Army, 1st Lieutenant, nurse, 71st Evacuation Hospital
Vietnam: 1969–70

The Unseen Enemy

George Veldman learned early on that danger could be hiding anywhere in South Vietnam.

Once, his unit came across a skull that turned out to be booby-trapped with a hand grenade. "That was something that you would, out of curiosity, want to pick up or move," he explains. "But you learned to not do that."

Veldman remembers the day several squads were separated in the mountainous terrain near Duc Pho. The first two squads radioed back that they had been hit by snipers. The medic was traveling with Veldman's squad, which was also pinned down by sniper fire. Veldman told the medic it was too dangerous to get to the wounded men and he should stay put.

"But the radio people down there were screaming bloody murder that they had to have the medic right away or somebody was going to die," Veldman remembers. "The medic was listening to that, and he just jumped up. He was shot and killed instantly."

After several hours, the snipers let up, and Veldman's men went back to get the bodies of two fallen Soldiers. Suspicious, they approached gingerly. Sure enough, the enemy had placed grenades underneath both bodies.

With the 27-man platoon now down to 12 who weren't either killed or wounded, the Soldiers set up perimeters to spend the night. The next morning they moved to a spot on top of a hill, where they would wait for a helicopter.

"Obviously we were tired and weary, so this big clearing looked like a good spot to spend the evening," says Veldman. As we moved up to it, one of the men in my squad, Lester, stepped on a booby trap. And I don't remember too much beyond that."

Lester lost part of his leg, and Veldman received shrapnel wounds on the whole left side of his body, requiring more than 400 stitches.

When the medevac choppers arrived, the pilots wouldn't land for fear there might be more hidden traps. "If they had hit a booby trap when they put those runners down, and flipped over the helicopter, it would have killed everybody up there. So they hovered right down above the ground as we were lifted up into the helicopter."

Veldman says his faith helped him keep going through so much uncertainty. "I knew there were a lot of prayers from my family going for me. I truly believe that you've got to have faith in something when you're in that environment, because you don't know what's going to happen at any second."

— George Veldman Jr.
U.S. Army, Sergeant, Company C, 11th Infantry Brigade (Light), 23rd Infantry Division
Vietnam: 1969

I Honor All of Them

He was so pumped with adrenaline, Robert Delsi didn't even feel the bullets rip through both of his legs. His Marine Corps platoon had just been ambushed in the Que Son Mountains in northern South Vietnam, and 19-year-old Delsi lay in the pouring monsoon rain, not calling for help so as not to draw more fire.

Eventually, some of his fellow Marines found him and carried him away from the firefight. Because the shooting was so intense, he and another wounded Marine waited for days before they could be medevaced out. Delsi could hear the automatic weapons fire and rocket-propelled grenades of the enemy surrounding him, and only U.S. naval gunfire and Air Force bombers kept them from being overrun.

"Every branch was trying to get us out of there," Delsi says. "To this day I honor all of them."

Finally, after three days, a Korean Marine dove in front of him, protecting him from enemy fire while a medical helicopter swooped down. When he asked about that Marine later, Delsi was told there were no Korean Marines in his unit.

And he wasn't the only "guardian angel" Delsi describes. After three days in mud and rain, his legs were swollen and infected. He was medevaced to the Naval Hospital in Guam, where doctors put both legs in casts and worried they'd have to amputate. After a week, Delsi asked a Navy corpsman if there was any way he could have a full-body shower, something he hadn't had since arriving in Vietnam. Delsi remembers the corpsman picking him up and carrying him into the shower. Red dirt and blood ran all over the corpsman's white uniform, Delsi recalls, but the man never complained. Later, when Delsi tried to find the man to thank him, nobody knew who he was.

It took months for Delsi to heal from his physical injuries, and much longer to recover from the emotional ones. He and his wife, Josie, organized "Patriots-Warriors to the Wall," a group of vets who ride motorcycles to the Vietnam Memorial in Washington, D.C. Along the way, they stop to visit families of fallen veterans.

Delsi has also found peace in making Native American warrior necklaces. He presents these necklaces and other gifts to veterans of all wars—and all branches of service.

— Robert S. Delsi
U.S. Marine Corps, Private First Class, infantry, Company A, 1st Battalion, 3rd Marines, 3rd Marine Division
Vietnam: 1969

Private First Class Raymond Rumpa, from St. Paul, Minnesota, walks past a burning hut of a Viet Cong base camp.
Photo courtesy of NARA.

Soldier guides a Huey helicopter onto landing zone south of Quang Tri.
Photo courtesy of NARA.

1970–Today
Coming to an End

As the new decade began, what had been covert U.S. bombing of NVA facilities in Cambodia turned into a land incursion, with President Richard Nixon requesting 150,000 additional Soldiers. Protests proliferated on college campuses.

As U.S. troops continued to sacrifice in Vietnam, Americans were still conflicted about the war, as summed up in two popular bumper stickers: "Give Peace a Chance" and "America: Love It or Leave It."

Meanwhile, the peace talks in Paris dragged on, and finally broke down at the end of 1972. Nixon ordered renewed bombing of North Vietnamese airfields and supply depots, and in early 1973, the U.S., RVN, DRV, and Viet Cong signed the Paris Peace Accords, establishing a cease-fire. American POWs were released, and the last U.S. combat Soldiers went back to what they called "the world." The war was officially over and the RVN was still intact, though reeling.

But two years later, in a violation of the Paris Peace Accords, the North Vietnamese Army swept through South Vietnam in a massive armored offensive. On April 29, 1975, as the NVA moved toward Saigon, U.S. Marines and Air Force helicopters airlifted nearly 7,000 South Vietnamese refugees to safety, the first wave of what eventually would be hundreds of thousands of refugees who resettled in the United States and other countries. A rocket attack added two final names to the list of U.S. Vietnam War dead: Marines Charles McMahon and Darwin Judge. The next day, the enemy took Saigon.

In total, 2.7 million American men and women served in Vietnam, stepping forward to accept duty to their country even in the face of increasing disdain and ultimately outright hostility from their anti-war peers. More than 58,000 were killed in action, and another 75,000 severely disabled. Yet more than 90 percent of Vietnam Veterans say they were glad they served, and over 70 percent say they would serve again if called.

Lieutenant Edward Ridgley, commanding officer, Company C, 3rd Battalion, 47th Infantry, 9th Infantry Division, calling in air strike and artillery on Viet Cong who have his company pinned down following an ambush west of Ban Tra village. *Photo courtesy of NARA.*

TIMELINE 1970–75

January 14, 1970
Diana Ross and the Supremes perform their last concert together, at the Frontier Hotel in Las Vegas.

March 25, 1970
Concorde makes first supersonic flight, traveling 700 mph.

March 31, 1970
U.S. Army brings murder charges against Captain Ernest L. Medina (later acquitted) for the massacre of Vietnamese civilians at My Lai in March 1968.

April 13, 1970
Apollo 13 is crippled by an on-board explosion that prevents a planned moon landing, but the three-man crew returns safely.

April 29, 1970
U.S. and South Vietnamese troops invade Cambodia.

May 4, 1970
Four college students at Kent State in Ohio are killed during a violent anti-war demonstration on campus.

May 8, 1970
The Beatles release their final album, *Let it Be*.

June 22, 1970
President Nixon signs the 26th Amendment, lowering the voting age to 18.

June 24, 1970
U.S. Senate votes to repeal Gulf of Tonkin Resolution.

June 29, 1970
U.S. troops pull out of Cambodia.

September 18, 1970
Rock star Jimi Hendrix dies of a drug overdose.

September 21, 1970
NFL Monday Night Football makes its debut on ABC.

February 8, 1971
South Vietnamese forces begin attacks on Ho Chi Minh Trail in Laos.

A UH-1 Huey medevacs an injured man about 12 miles south of the DMZ.
Photo courtesy of NARA.

The Last Big Battle

Company D had been fighting for months in the hills and jungles of South Vietnam's A Shau Valley, and now it was the end of July 1970. They had just been sent back to Camp Evans for training and a few days of rest when they heard, "All D Company, report back immediately!"

"We've got a terrible mission, men," their commanding officer told them, explaining that Alpha Company of the 2nd Battalion had been ambushed by the North Vietnamese Army. "We're going out there to rescue them."

Alpha had been defending Fire Support Base (FSB) Ripcord, which the U.S. Army had established earlier that spring on a strategic hill above the A Shau Valley. In rotation with other companies, Alpha had taken turns defending the firebase and the artillery battery there and patrolling the jungle. The goal of FSB Ripcord was to halt movement of troops and supplies on the Ho Chi Minh Trail; the goal of the NVA was to totally eliminate the American presence on or around the hill.

As Delta Company prepared to load up in helicopters, the sergeant told Gilbert, "And by the way, Freddie, you're going to be our point man." Gilbert had been one of the company's point men ever since he'd been promoted from M60 ammunition bearer. It turned out that Gilbert, a country boy who knew how to stalk deer, was a natural.

The helicopters arrived in the A Shau Valley, but had to turn back because every landing zone was under intense fire. *I know these guys can't be out there much longer, leaking blood,* Gilbert remembers thinking about the Soldiers under siege.

At first light the next day, the helicopters lined up again. This time, Gilbert's group was able to land, but the squad leader was shot as soon as he jumped off the Huey. Gilbert jumped down behind him and started running point, jumping over the carnage of dead Soldiers. It took 40 minutes of running to reach the beleaguered men of Alpha Company. Company D drove off the NVA, then blew down trees to create a new LZ, and helped the injured Soldiers onto each Huey as it landed.

"It was the last major battle of the Vietnam War," Gilbert says. But there was no media coverage of Ripcord because of the firestorm over a previous battle—Hamburger Hill, "the one everybody knows," Gilbert says. "But I got 248 men chiseled on the [Vietnam] Wall from a battle called Ripcord, and that's where my heart is."

— **Freddie Gilbert**
   U.S. Army, Specialist 4, Company D, 2nd Battalion, 506th Infantry (Airborne), 101st Airborne Division
   Vietnam: 1969–70

Based on an interview with Freddie Gilbert, Veterans History Project, Grand Valley State University.

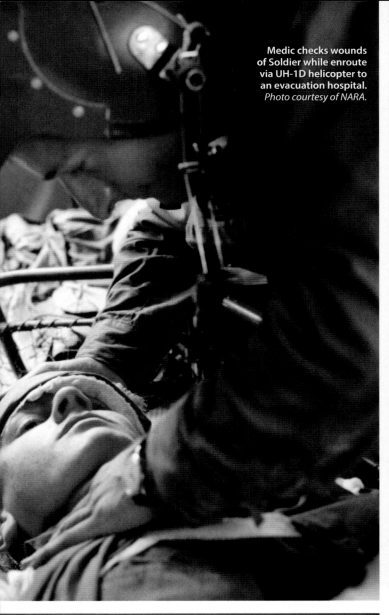

Medic checks wounds of Soldier while enroute via UH-1D helicopter to an evacuation hospital. *Photo courtesy of NARA.*

U.S. Air Force F-4C Phantom fighter bombers refuel from a KC-135 tanker prior to making a strike against targets in North Vietnam. The Phantoms are fully loaded with 750-pound bombs and rockets. *Photo courtesy of NARA.*

March 29, 1971
U.S. Army court-martials Lieutenant William Calley for his part in the My Lai massacre.

June 28, 1971
U.S. Supreme Court reverses boxer Muhammad Ali's 1967 conviction for refusing induction into the U.S. Army on religious grounds.

April 29, 1971
Total U.S. deaths in Vietnam surpass 45,000.

June 13, 1971
The New York Times publishes the Pentagon Papers, a secret Department of Defense study of U.S. involvement in Vietnam.

December 17, 1971
U.S. troop levels drop to 156,800.

February 22, 1972
President Nixon visits China, opening new possibilities for trade.

March 30, 1972
North Vietnam launches Easter Offensive.

May 15, 1972
The headquarters for the U.S. Army in Vietnam is decommissioned.

September 5, 1972
Arab commandos kill 11 Israeli Olympic team members at the 1972 Summer Olympics in Munich, Germany.

September 17, 1972
TV program *M*A*S*H* premieres on CBS.

September 20, 1972
Tennis star Billie Jean King beats Bobby Riggs in "Battle of the Sexes" tennis match.

November 7, 1972
President Nixon is re-elected to a second term in one of the biggest landslides in U.S. history.

December 29, 1972
President Nixon halts bombing in Vietnam.

A Couple of Old Grizzled Fighter Pilots

The first time Dan Cherry saw Nguyen Hong My, the North Vietnamese fighter pilot was falling from the sky. Just seconds before, the two men had been in a dizzying dogfight. Cherry had fired a Sparrow missile from his F-4D Phantom fighter and hit the right wing of Hong My's MiG-21, causing it to spin toward the ground. The enemy pilot had ejected and was now suspended from a red-and-white parachute less than 100 yards away.

It was April 1972, and the U.S. had just resumed bombing raids on North Vietnam after a four-year hiatus.

"A fighter pilot's war is very impersonal," Cherry says. "Your objective is to disable or destroy the airplane; you're thinking about it as a machine, not about the human inside." For the first time in his two tours flying over Vietnam, however, Cherry now wondered about the man he had shot down. Who was he? Did he have a family? Did he survive? But this was war, and soon Cherry was back in the air on another mission.

Then 35 years went by. And—very long story short—on a trip to Ohio, Cherry and some buddies discovered that the same Phantom he had flown in that historic dogfight, plane #66-7550, was on display in front of a VFW club. The plane looked pretty bedraggled: the tires were flat, the fins had rusted through, and there were bird droppings everywhere. But the side of the plane still bore his name, though faded, along with the red "victory star" commemorating a MiG shot down in battle.

Cherry and his friends raised money to have the plane restored, then created Aviation Heritage Park, a museum and educational center in Cherry's hometown of Bowling Green, Kentucky. And he began to wonder again if that MiG pilot had survived. One query led to another, and in April 2008, Cherry found himself in Ho Chi Minh City (formerly Saigon), Vietnam, on the set of "The Separation Never Seems to Have Existed," a live Vietnamese TV show dedicated to spectacular reunions. Cherry was invited onto the stage, and then a man stepped out from behind a partition.

"Our eyes locked into each other," Cherry recalls. In Vietnamese, Nguyen Hong My said, "Welcome to my country. I'm glad to see you're in good health. I hope we can be friends." Pictures of their children and grandchildren flashed on the screen behind them, and both men were overcome with emotion. "Here we were," Cherry remembers, "a couple of old grizzled fighter pilots crying our eyes out on television."

After the show, Hong My invited Cherry to Hanoi to have dinner with his family, including his 1-year-old grandson. "When we approached the front door, his son came out with the stroller and he handed me his grandson. I knew then we were going to be friends. I mean real friends."

Longer story short: The circle of friendship widened. Hong My wondered what had happened to the pilot of a plane he himself had shot down, and eventually Cherry was able to track down pilot John Stiles and helicopter crewman Bob Noble, who rescued Stiles from the jungle. On a visit to America, Hong My met Stiles and Noble.

Cherry later wrote a book about all this; he called it *My Enemy My Friend*. Hong My likes the book but not the title. "I don't think we were ever really enemies," he told Cherry. "We were just Soldiers."

— **Dan Cherry**
U.S. Air Force, Major, fighter pilot
Thailand: 1967–68; 1971–72

Cherry was later promoted to Brigadier General and retired after 29 years of service.

Guided by long-range radar (LORAN), a flight of three F-4D Phantom II fighters and three A-7 Corsair II attack aircraft drop bombs during a strike mission. *Photo courtesy of NARA.*

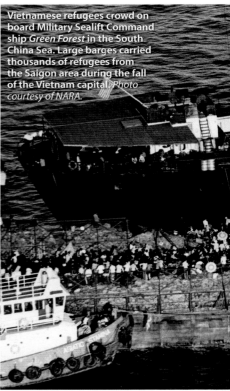

Vietnamese refugees crowd on board Military Sealift Command ship *Green Forest* in the South China Sea. Large barges carried thousands of refugees from the Saigon area during the fall of the Vietnam capital. *Photo courtesy of NARA.*

Hanoi Hannah

A courier for Air Combat Intelligence (ACI) during the Vietnam War, Walter Washington often traveled on low-flying helicopters to deliver classified information to military officials from Japan to Vietnam. He carried with him top-secret map overlays in transparency format—and nothing else. ACI wanted to limit the amount of information the enemy could obtain, in case he was ever captured.

Thankfully, Washington wasn't ever captured, but ACI's attempts to keep its information secret weren't always successful. In fact, Washington says, the most frustrating aspect of dealing with top-secret information involved the infamous "Hanoi Hannah" and her propaganda radio shows.

"We had a lot of Top Secret materials that would be declassified from Top Secret to Secret to Confidential," he explains. "Once it got to Confidential, we would burn it on Thursdays, under armed guard." Despite these carefully guarded "burn Thursdays," the same information would often appear within days on the front page of the *Japan Times* and in Hanoi Hannah broadcasts, sometimes even publicly identifying U.S. troop positions.

"It seems there were leaks in our system that allowed the Japanese media and Hanoi Hannah to have information about exactly what was going on with us. It provided a lot of discomfort and insecurity," Washington recalls.

In 1991, Don North interviewed Hanoi Hanna for "The Sixties Project," sponsored by the University of Virginia. Hannah, whose real name is Trinh Thi Ngo, stated that the main source of her intelligence actually came from the *Stars and Stripes,* which her group read daily, as well as *Newsweek* and *Time* magazines. She said they also intercepted the AP and UPI wires.

Although much of her information was derived from public information, "her broadcasts were designed to chill and frighten the young troops, and she was definitely effective," Washington observes.

— **Walter Washington**
  U.S. Marine Corps, Staff Sergeant,
  1st Marine Air Wing, Fleet Marine Force, Pacific
  Vietnam: 1971–72

January 27, 1973
The U.S. and North Vietnam sign the Paris Peace Accords.

February 12, 1973
Operation Homecoming begins with the release of 591 American POWs from Hanoi.

March 29, 1973
The last remaining U.S. troops withdraw from Vietnam.

April 1, 1973
Captain Robert White, last known American POW, is released.

January 15, 1974
TV program *Happy Days* premieres on ABC.

April 8, 1974
Hank Aaron hits 715th home run, surpassing record set by Babe Ruth.

May 9, 1974
U.S. Congress begins impeachment proceedings against President Nixon for the Watergate scandal.

August 9, 1974
President Nixon resigns; Gerald Ford is sworn in as 38th U.S. president.

April 23, 1975
President Ford, speaking in New Orleans, calls the Vietnam War "finished."

April 30, 1975
At 8:35 a.m., the last Americans depart Saigon, ending U.S. presence in Vietnam. North Vietnamese troops pour into Saigon. The war is over.

November 13, 1982
Vietnam Veterans Memorial is dedicated in Washington, D.C.

From Leeches to Linens

The jungles and the deltas, the grunts and the generals, the firefights and the typewriters: not everybody who served in South Vietnam experienced the same war. But Larry Buehner came as close as anyone to seeing the big picture.

For the first seven months in Vietnam, Buehner was a scout dog-handler with the 37th IPSD (Infantry Platoon Scout Dog), humping 80 pounds of gear, food, and water in the field. He and Cali, his German Shepherd, would usually be walking point, the first in line out on patrol, on the alert for punji pits* and trip wires and the smell of enemy soldiers. "Everyday life was picking leeches off yourself, off the dog," he says.

He remembers a particular search-and-destroy mission through the jungle when Cali gave an alert that signaled danger, but Buehner's lieutenant told the platoon to push on. "Against my better judgment, we moved forward, when again Cali gave another strong alert," he remembers. This time, Buehner refused to go any farther.

After radio transmission back and forth with another platoon also reporting "movement," it became clear that Buehner's platoon had nearly walked into a friendly-fire ambush. "Cali saved us that day from someone's error. If it weren't for the military working dogs in Vietnam, a whole lot more of us would never have made it home. I owe my life to a dog."

The second half of Buehner's tour was very different. He served as a photographer with the Public Information Office, taking pictures for the brigade newspaper and the *Stars and Stripes*, the government-authorized military newspaper. Sometimes he photographed combat assaults; sometimes he covered the top brass.

The officers stationed in the rear—"the guys who always wore starched fatigues, who had shiny, spit-polished boots all the time"—ate on china plates with gold trim, he reports. They drank out of water goblets and had linen tablecloths. Buehner wondered what the Soldiers out in the field, the guys he served with for the first half of his tour—the guys whose socks were soggy, out in the jungle—would think of all that elegance.

— Larry Buehner
U.S. Army, Specialist 5th Class, scout dog-handler, 37th Infantry Platoon Scout Dog (IPSD); photographer, 3rd Brigade PIO; 1st Air Cavalry Division
Vietnam: 1970–71

*A punji stake is a one- or two-foot-long sharpened piece of bamboo or long metal spike. A punji pit is a hole camouflaged by leaves, with punji stakes protruding out of the sides and the bottom. When someone stepped into a punji pit, he was initially injured by the vertical stakes in the base; the lateral downward-positioned stakes made his attempts to get out extremely difficult. Additionally, the points of the punji sticks were often smeared with poison or animal excreta.

Army Specialist Larry Buehner and his dog, Cali, await a chopper ride back to the firebase. *Photo courtesy of Larry Buehner.*

USS *Ranger* (CVA-61) flight deck crewmen take a break in operations near an A-1 Skyraider. *Photo courtesy of NARA.*

A bus transports troops leaving Vietnam. *Photo courtesy of NARA.*

Tie a Yellow Ribbon

The man's voice on the other end of the line was familiar. Still, Tony Orlando thought it was a joke when the man said, "Tony, this is Bob Yellow Ribbon Hope."

"This is Frank I-don't-believe-you Sinatra," Orlando shot back.

It was the spring of 1973, and the group "Tony Orlando and Dawn" had just released a song about a guy who was returning home by bus after "three long years" away. Would his love still want him? the guy asked the bus driver. Would she tie a yellow ribbon 'round the old oak tree to show him she wanted him home?

Bob Hope was arguably the most famous comedian in America then, and he had made scores of USO trips to entertain the troops during World War II, the Korean War, and the Vietnam War. He wanted Orlando to sing "Tie a Yellow Ribbon" at the Cotton Bowl, at a performance to welcome home American prisoners of war from Vietnam, Laos, and Cambodia. "That line from your song—'I'm coming home, I've done my time'—is perfect to open the show," Hope said.

"I don't want to disappoint you," Orlando answered, "but no one knows that song."

And Hope replied: "By the time you come down to the Cotton Bowl, that song will be No. 1." And two weeks later it was. Eventually, in fact, it was the Billboard #1 song of 1973.

There were 70,000 people in the audience at the POW homecoming at the Cotton Bowl that April. "I see Bob Hope smiling at me with a big wink," Orlando remembers. "We finish the song and there was one former POW on the front row who wasn't getting into the song, wasn't clapping. I asked Bob Hope, 'Do you think I offended him?' And he said, 'Go ask him.'" So Orlando did.

The POW apologized. "I'm sorry," he said. "My shoulders have been pulled out and I can't clap, but what you can't see is my big toe tapping along." The former POW introduced himself: "I'm John McCain."

When Orlando performed that year at the Copacabana, he flew 10 of the returned POWs to New York City, and convinced the management to dress up the nightclub with yellow ribbons. It was the start of 40 years of shows that have raised hundreds of millions of dollars for veterans' causes.

In 2013, on the 40th anniversary of the POWs' homecoming, Orlando donated his "Tie a Yellow Ribbon" gold record to the Nixon Presidential Library and Museum. By then, yellow ribbons had become symbols of hope and homecoming for every Soldier who goes off to war.

— **Tony Orlando**
Singer, USO tours to Vietnam

The Other Enemy

Like many Soldiers in South Vietnam, Kimball Phillips battled a dangerous, deceptive, and deadly enemy—heroin. Produced in Southeast Asia's poppy-growing "Golden Triangle," the drug was pure, uncut, inexpensive, and easy to obtain.

"Many of the Soldiers resorted to either alcohol or drugs, perhaps due to stress, low morale, being away in a strange country, or boredom," he explains. "In my unit, I know at least 20 percent had a problem with heroin."

Dabbling in drugs brought Phillips to South Vietnam in the first place. As a teen, he and his friends experimented with marijuana. Nearing high school graduation, they tried selling pot for quick money; Phillips sold some to an old school chum, who happened to be working undercover for the police. Phillips landed in jail.

"Since I had no prior record, and they needed Soldiers for Vietnam, it was either a long probation, or enlist in the Army," Phillips says. "I chose the Army. It was a place to get away from what had happened, and it seemed like a big adventure."

He trained as a helicopter crew chief/door gunner on the UH-1 (Huey) helicopter. Stationed at Vinh Long Army Airfield, he went out on a few combat assaults. But since he could type, he was asked to become a clerk typist. "I wasn't happy about that, until I saw some of our helicopters come back with casualties and battle damage," he says.

But his brush with the law hadn't cured his urge for marijuana. He soon became part of the pot-smoking Soldiers known as "potheads."

"At the time, I believed marijuana made me cope and operate better, physically and mentally," Phillips says. "Now I realize the real danger of any drug is it makes you feel superior. It skews your thinking, and you're trying to justify using it."

The first time he tried to buy marijuana, he sought out an old Vietnamese man in town. "Instead of pot," Phillips recalls, "the old man brought me back the drug that was easiest to get. Wrapped up in a newspaper were ten plastic one-inch capsules filled with pure, uncut heroin. It could have sold on the streets of New York for $12,000–$15,000. He offered it to me for $35." But Phillips declined.

"I was a small-town boy from Utah," he explains. "I didn't realize what I held in my hands. I told him I wanted marijuana." So the man brought back a big bag of pot.

"It was very cheap and very potent," Phillips said. "A few months later, a friend who smoked heroin got me to try it. Heroin gives you a euphoric high that separates you from all your troubles. But I didn't realize the physical effects." He refrained from injecting it as some Soldiers did. "That would really put you out, and I couldn't do my job or guard duty if I was impaired."

There were rumors the Army was conducting unit sweeps with unannounced drug tests. "But we didn't believe it would happen to us," Phillips says. One Sunday morning, military police suddenly showed up and made everyone line up for urine tests. "I was only smoking it [heroin] sporadically, not every day, so I didn't think the urinalysis would pick it up. But it did. My platoon sergeant was shocked."

Those Soldiers testing positive were taken to a detox center. "It was like a minimum-security prison with a barbed-wire fence and guards," Phillips recalls. "It was a Quonset hut with about 25 cots, a lot of male nurses, and a few doctors. Everybody went cold turkey. They didn't give us any medications, just a lot of fluids to try to flush out our system."

The withdrawal was agonizing. "I was shaking, and I don't think I slept for four days," he says. "But that was mild from what I saw around me. People were moaning, begging for help. It was a lot worse because of the purity of the drug. Heroin takes away your appetite, and some of the guys were skin and bones, like they had come out of a concentration camp."

The detoxing Soldiers had to attend a few classes about the harmful effects of drugs, and were then released once their urinalysis tests were clean. "I stayed six days," Phillips says. "It was a harsh lesson on high-powered drugs. By the time I got out of the service, I had grown up and realized they weren't for me. I made a change of friends, and we had jobs and families."

Like many Vietnam Veterans, Phillips never relapsed into past habits after coming home, but he believes his drug use hindered him from reaching his potential. "I wasted that time, when I could have gotten serious about a career earlier in life," he says. "Yeah, I had a good time temporarily, but it was short-lived."

— **Kimball Franklin Phillips**
   U.S. Army, Specialist 4, helicopter crew chief/door gunner, clerk, 175th Assault Helicopter Company, 214th Aviation Battalion
   Vietnam: 1971–72

Sheer Boredom and Pure Panic

"Many hours of sheer boredom and sudden minutes of pure panic" is how Edward Timm described his job. Timm was a watchstander for the LORAN (Long-Range Aids to Navigation) monitor station on the Udorn Royal Thai Air Force Base in Thailand.

The general public is unaware that the Coast Guard played this critical part in the Vietnam War, Timm says. "LORAN was the system of navigation used by all other services," he explains, "and ours was the monitor station that kept the other ones on time and in tolerance. If it was off, someone could end up flying into obstacles—like the side of a mountain."

Two or three times when he was on graveyard watch, Timm got phone calls in the middle of the night from a colonel or general of a tactical fighter wing in South Vietnam, asking how the LORAN was tracking that night. "There must have been a mission planned and they needed to make sure the navigation system was working properly," he says. "In fact, I heard about two instances where F-4s were lost because of the out-of-tolerance and instability our chain was having with one of the stations."

Maintaining the LORAN system was considered direct combat support. "They told us we maintained the tightest tolerances of any LORAN chain in the world," said Timm. "As the years have gone by, I've come to realize that what we did was more important than what I thought at the time. I'm proud that I served in the Coast Guard. I wouldn't trade the experience for anything, but I don't know if I'd want to do it again."

— Edward Timm
U.S. Coast Guard, Seaman, LORAN watchstander
Thailand: 1970–71

Explosive Ordnance Demolition (EOD) personnnel conduct a controlled burn of unwanted ammunition near Saigon.
Official U.S. Navy photo courtesy of Russell A. Elder.

A U.S. Coast Guard WPB cutter takes part in Operation Market Time, the anti-infiltration patrol of South Vietnam's 1,200-mile coastline.
Photo courtesy of DoD.

One of Those Days

Being a scout helicopter pilot led to some tense times for Jim Wise while flying in Vietnam; essentially, he was enemy bait. "My job over there was to fly as low and slow as I dared, to try to get shot at—and it worked," he explains.

The scouts would hover at treetop level, blow the tall elephant grass aside, and look under trees for signs of the elusive enemy. "When there were buildings out in the middle of nowhere," Wise recalls, "we were known to hook the toes of our skids underneath the roofs and lift them up, to see if anybody's home, because what's a building doing out in the middle of nowhere?"

AH-1 Cobra helicopters acted as cover ships. When a scout helicopter was fired on, the scouts would kick out a smoke grenade to mark the location. The Cobras would look for the smoke and come in firing. "The Cobras initially aimed directly at us, to put the rockets underneath and suppress anyone shooting at us," Wise says.

They didn't get shot at every day. They might fly six or seven hours, he says, without seeing anything unusual—"but then there were moments of stark terror." One day when Wise received heavy fire, Billy Joe Miller was flying overhead in the Cobra. "Billy Joe told me he had never heard my voice go so high," Wise recalls. Another day, Wise was taking enemy fire and kicked out a smoke grenade, when he heard Dan Sloan, the Cobra pilot covering him, say over the radio, "Three-four's inbound... Oh shit!"

"I didn't know what that meant, but I knew it wasn't good," says Wise. "I pulled in more power and kept on going. Our Cobras' inboard rocket pods had either a 10- or 17-pound warhead of high explosives. The outboard pods would carry flechette rockets—they contained 2,200 little nails about an inch-and-a-half long, with little fins on them. They were so sharp you could barely tap your finger on one and you'd puncture your skin."

Sloan had forgotten where he had the pod selector switch set, and it was on "outboard" instead of "inboard." He realized too late that he had mistakenly fired flechettes in Wise's direction—hence the expletive. "All those little nails were fanning out to cover the area he aimed at," he explains. "I didn't get hit with them, but heard the swoosh as the nails went by over the sound of the rotor system. That got my attention."

It was a high-stress moment. "But Dan bought me an extra beer that night," he says. "You just had those kinds of days."

— James Frederick Wise
U.S. Army, Warrant Officer 1, pilot, Troop A, 7th Squadron, 17th Cavalry, 10th Aviation Battalion (Combat)
Vietnam: 1971–72

Specialist 4 Jim Messengill, Team 22, Company H (Ranger), 75th Infantry, reports contact with enemy forces during a patrol along the Dong Nai River. *Photo courtesy of NARA.*

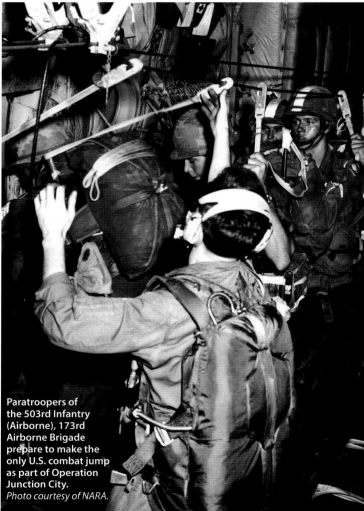

Paratroopers of the 503rd Infantry (Airborne), 173rd Airborne Brigade prepare to make the only U.S. combat jump as part of Operation Junction City. *Photo courtesy of NARA.*

An M113 Armored Personnel Carrier and an M551 Sheridan Armored Reconnaissance Assault Vehicle from Troop A, 3rd Squadron, 4th Cavalry, 25th Infantry Division take positions along the perimeter fence of LZ Hampton. *Photo courtesy of NARA.*

A Music High

On the battlefield, say veterans who served in the Vietnam War, "everyone bleeds the same color." But that inclusivity didn't always persist once Soldiers got back to the rear area base, where black and white Americans often kept themselves segregated, and racial tensions sometimes escalated into fist fights. Which is why DJ LeRoy Thomas started playing what he calls "integrated music."

Thomas was stationed at bases in Thailand, in two tours that ranged from 1970 to 1973. His official duties included maintaining the equipment that helped arm the U.S. fighter planes at Ubon and Korat air bases.* But when he returned for his second tour, he volunteered his time as DJ for the bases' enlisted club and as host of a weekly radio show over the Armed Forces Thailand Network (AFTN)** in Korat.

The slogan of the Armed Forces Radio during the Vietnam War was "the voice of home," but Thomas noticed that the music was more likely to have a country twang than a soul vibe, and the jukeboxes on the bases had a John Denver-Marvin Gaye ratio of about 10 to 1. So Thomas initially tried to even things out with a radio show of solid soul, but he soon began spinning a more integrated mix to appeal to a wider range of listeners. The program was called the "Sweet Daddy Lee Show."

His typical sign-on: "Owwwww! Good goo galar moo gah! You're sitting in with the soul sugah, Sweet Daddy Lee. Sockin' the pure unadulterated soul to ya! Gots ta do it 'cause I'm used to it!"

Thomas would play a mix of songs designed to include something for everybody—classic rock, pop, country, and soul. And he would seamlessly segue from one to another in a way that blurred the differences. "I would take the drum solo from Iron Butterfly's *In-a-Gadda-Da-Vida*, which most Caucasians knew, and mix it with Santana's *Soul Sacrifice*, which most brothers knew, or something from AWB [the Average White Band] and mix it with a James Brown beat," Thomas explains.

The result, he says, was that "everyone is experiencing each other in a new way and enjoying each other's music while they're drinking, joking, and laughing together. Whenever I had a show, I never had anyone fight. I kept people on a music high… the magic carpet ride."

— **LeRoy Thomas**
    U.S. Air Force, Senior Master Sergeant, Armed Forces Thailand Network
    Thailand: 1970–73

During the Vietnam War, about 80% of all USAF air strikes against North Vietnam—as well as attacks against Communist forces in Laos, South Vietnam, and Cambodia—originated from at least six air bases in Thailand.

**The U.S. Air Force started Armed Forces Thailand Network (AFTN) in 1966. The network had stations on the air at six bases and more than 20 satellite stations that rebroadcast the primary stations.*

Marine Staff Sergeant Ermalinda Salazar was nominated for the 1970 Unsung Heroine Award presented annually by the Ladies Auxiliary to the VFW. She helped the children of the St. Vincent de Paul Orphanage in Vietnam in her off-duty hours. *Photo courtesy of NARA.*

Welcome Home

As a combat medic in Vietnam, José G. Ramos worked hard to save a lot of his comrades. Years later, he's still trying to save them.

When Ramos joined the Army, he was an inexperienced 17-year-old, and knowing how to put on a Band-Aid was the extent of his medical skills. But he learned quickly in combat. "I saw a lot of trauma and I lost a few men, but it was the greatest job in the world," he says. "The infantry guys were amazing. Every morning they would get up and grab their weapons and chow, and walk into the jungle. What a powerful group of men."

But instead of a warm welcome home from a grateful nation, many of these amazing veterans were greeted with scorn.

"I came back with survivor's guilt and post-traumatic stress disorder," Ramos says. "We came home confused. We didn't know we were considered the bad guy. I'm sitting in a bar and seeing all this talk of baby-killers on the TV screen, and I suddenly realized it was me they were talking about. It wasn't long before I started telling people I never went to Vietnam, I went to Germany."

Ramos fell into drugs and alcohol, and attempted suicide. He spent several months in VA hospitals. "From the PTSD group that I had only been in for five years, four members had died," he says. "One was self-inflicted, and another had a bad drinking problem. I realized that at our ages, we were going to go pretty fast, and I wanted to do something about that."

Rather than give up, Ramos became determined to change things. In 2002, he organized the first Welcome Home Vietnam Veterans Day in his hometown of Whittier, California. Then in 2004, he rode his bicycle from Whittier to Washington, D.C., lobbying for a national day to give Vietnam vets the welcome they never got when they came home. His efforts paid off. In 2009, Governor Arnold Schwarzenegger, then governor of California, signed a bill proclaiming "Welcome Home Vietnam Veterans Day" at the Twentynine Palms Marine Base, with Ramos standing by his side.

Governor Schwarzenegger praised Ramos's tenacity. "He comes up with the idea and then he's like a tick," the governor said. "He hangs on you and he fights and he fights and he fights until he gets it done."

In 2012, Ramos's idea went all the way to the top. President Barack Obama signed a proclamation designating March 29 as Vietnam Veterans Day.

"I know it has made a difference in peoples' lives," he says. "We held an event here in 2008, and 5,000 people came." He described the events as a "living memorial for the living, not the fallen. We have hot dogs and burgers and we tell little stories about the war, and we have fun. Every year we find a veteran who did something exceptional and never got recognized, and we have Congressional people come and give medals and awards." Veterans across the country have expressed thanks to Ramos for finally being recognized and supported for their service.

Ramos is also working to establish a Vietnam Veterans "living memorial" that focuses on teaching the arts. "Everywhere you go, there are memorials to the fallen, but this is something for the living," he says. In 1998, he returned to Vietnam for a 1,200-mile bicycle ride from Hanoi to Ho Chi Minh City (formerly Saigon) with a group of veterans. The trip was turned into a Kartemquin Films documentary, *Vietnam, Long Time Coming*.

Ramos credits Vietnam Veterans for changing the way America treats all its veterans. "They came home, went back to work, mowed their lawns, and raised beautiful children and grandchildren," he said. "And when the next war started, they went to the airports to welcome those troops home. Never again will one veteran leave another veteran behind. I'm so proud of them, because that's exactly what we need."

— **José G. Ramos**
  U.S. Army, Specialist 4, medic, 3rd Battalion, 506th Infantry (Airborne), 101st Airborne Division
  Vietnam: 1965–68

Cornelius Harrington Yates III stands near his helicopter. *Photo courtesy of Skip Yates.*

Operation End Sweep

By the time the United States sat down to negotiate with Hanoi at the Paris Peace Accords in January 1973, the Americans had laid down over 11,000 mines in North Vietnamese waters. As a bargaining point in the peace talks, the U.S. offered to neutralize the mines, partly in exchange for a return of U.S. prisoners of war.

But once the ceasefire agreement was signed, the Mine-Clearing Protocol was much broader and more demanding than expected. "Operation End Sweep"—removing, deactivating, or destroying all 11,000 mines in Haiphong Harbor and other coastal and inland waterways—was a painstakingly tedious and lengthy process. U.S. Navy Commander Cornelius "Skip" Harrington Yates III returned to Vietnam to oversee the operation.

Yates had first been in Vietnam from 1967 to 1969 as part of a helicopter combat support squadron. When he returned in 1973, he joined three helicopter mine countermeasures squadrons (HMCS) manned by the Navy and Marine Corps. Ocean minesweepers, amphibious ships, and other parts of the U.S. naval fleet conducted most of Operation End Sweep, but the helicopters could do the sweeping an estimated three to six times faster.

"I made history," Yates says. "I was the first pilot to blow up a mine from a helicopter, towing a MK-105 hydrofoil (magnetic) minesweeping sled." Towed at speeds up to 25 knots, the foil-supported generator streamed a standard magnetic tail astern. A photo of this event made the front page of *The New York Times*, although an admiral "got the credit," Yates says.

Yates' minesweeping operation happened to be in the same area where he had once been a combat rescue pilot—the same place where he had been shot at countless times. So he had a different perspective than his co-pilot during Operation End Sweep.

"We would fly up the river, and my co-pilot would wave at the people. And I would give them the finger. So, different generation."

— **Cornelius "Skip" Harrington Yates III**
  U.S. Navy, Commander, helicopter pilot, Helicopter Mine Countermeasures Squadron 12
  Vietnam: 1967–69, 1973

UH-1D helicopters from the 227th Assault Helicopter Battalion return to Quan Loi after carrying members of the 7th Cavalry on an assault mission. *Photo courtesy of NARA.*

Better Than a Court Martial

On a Vietnam War mission in the spring of 1972, Air Force Major James R. Anderson and his AC-130 gunship crew from the 16th Special Operations Squadron played a key role in the defense of a South Vietnamese Army firebase. Air Force superiors, however, were not happy that Anderson had broken rules to do it.

But there was no time to do things by the book, Anderson explains. It was the enemy's big Easter Offensive. North Vietnamese Army units had surrounded the South Vietnamese base at Dak To, and 20 to 30 U.S. Army advisors were trapped there. To help, U.S. Air Force B-52 bombers were enroute from a base on Guam to make an air strike. Safety rules, however, required the low-flying AC-130 to stay 25 kilometers away from the bombing target.

"When we came back to Dak To," Anderson recalls, "the American advisors begged us not to leave them again. When we had left before, enemy armor moved in and worked them over badly. We promised we wouldn't leave again. But it wasn't 20 minutes later we were ordered to clear away from the target. We declined."

Anderson and his pilot offered to fly an orbit around the area, but would not leave entirely. Consequently, the B-52s were refused final permission to bomb, and eventually left for a secondary target. "Strategic Air Command* was very angry," Anderson says.

The AC-130 was running low on fuel, but the crew was unwilling to abandon the besieged men below. "My crew decided we would use our 'get-home fuel' to stay on target, because we knew there was a base nearby at Pleiku," Anderson says. "We were on fumes when we glided into Pleiku. With strong enemy forces on the offensive in the area, the base commander was not happy to see us." The commander did not want the plane, "a mortar magnet," near his fuel station. He gave Anderson's crew just enough fuel to get back to Ubon Air Base in Thailand.

"But as soon as we were airborne, we could hear Army guys on the radio yelling they were under attack," Anderson says. So the gunship flew back into the battle. "About that time, we got a call from the Navy," he recalls. "They said, 'We don't know why the Air Force won't help you, but we've been listening to you all night.'" Anderson's plane took out one enemy tank, and Navy planes blew up a second. Army helicopters were sent to evacuate the U.S. advisors, but one helicopter crashed with advisors aboard.

With the Navy handily breaking up the attack by enemy armor, Anderson and crew flew back to their base. Two messages were waiting for them. "One was from the gunship desk at Blue Chip [7th Air Force headquarters in Saigon] saying, 'Way to hang in there, guys,'" Anderson remembers. "The other was from the B-52 desk at Blue Chip, wanting to know the name, rank, and serial number of the senior officer and table navigator—which was me—and the pilot." Soon after, Air Force headquarters issued orders for Anderson and the pilot to report immediately to Saigon for Article 32 proceedings. Those measures, Anderson well knew, were preparations for a court martial.

Fortunately, the Army advisors in the crashed helicopter survived and hiked out of the jungle. They told Air Force officials how Anderson and his crew had saved their lives. "The next thing we knew, our entire crew was recommended for Silver Stars," Anderson recalls. "And a Silver Star is better than a court martial."

— James R. Anderson
U.S. Air Force, Major, navigator, 16th Special Operations Squadron, 8th Tactical Fighter Wing
Vietnam: 1968–69, 1971–72

Strategic Air Command (SAC) controlled all U.S. Air Force B-52s.

Members of Battery A, 7th Battalion, 8th Artillery read the news of the Apollo 13 splashdown in the April 1970 edition of the *Stars and Stripes*. *Photo courtesy of NARA.*

Marines riding atop an M-48 tank cover their ears as the 90mm gun fires during a road sweep. *Photo courtesy of NARA.*

Honoring the Heroes

"Don't try to be a hero," Waldo Fisher said when his son shipped off to the Iraq War in 2005. "I already have enough medals for both of us."

During the Vietnam War, Fisher had plenty of chances to earn medals in the war-torn A Shau Valley. For most of his tour he walked point, the lead Soldier in the patrols, most likely to be the first to face a firefight, ambush, or booby trap.

"I was born and raised in Arkansas," Fisher explains. "I had wandered through the woods and hunted all my life, so I felt more at ease doing it than counting on one of the guys from the North or the East. You get almost an instinctive feeling when something's not right, and the hair stands up on the back of your neck."

Fisher got that feeling one day in December 1970 when he was walking point with his platoon and came to an unusual bare spot, where napalm had hit and burned up the vegetation. "I told everyone to stop, because something felt wrong. But when I went back to tell everyone about it, my lieutenant kept going. He walked out in the area and got blown away by an ambush. They had been waiting for us."

Fisher quickly got everyone in a defensive position, and instead of staying to fight, the enemy ran. Fisher was given a Silver Star for his actions. His lieutenant, taken by a medevac helicopter to a hospital, lost a leg but survived.

Feeling like "an old man" when he came home from the war at age 22, Fisher earned a B.S. degree and began coaching and teaching civics in Arkansas.

In 1989, he started an annual Memorial Day tradition to have his students decorate the Fort Smith National Cemetery, placing up to 14,000 flags on the graves of fallen service members. Students looked forward to it all year, and it motivated them to avoid detentions that would keep them from being chosen.

"I'm a firm believer that all the education is not out of a book," Fisher says. "They all knew that I felt very honored to be a veteran, so they did it out of respect for me. When they see all those white tombstones for the first time, it takes them by surprise to realize what these men and women went through for them. It teaches them about citizenship and respect."

Then, a few days before Memorial Day in 2005, students watched as uniformed Soldiers came to the school to give their teacher the tragic news that his son Dustin had been killed in a car bombing in Baghdad.

His son's death brought Fisher's long-buried feelings from the Vietnam War to the surface. "I had convinced myself there was nothing wrong, that Vietnam hadn't affected me. But Dustin's death destroyed my life, to be honest. I just broke down."

He taught one more year, and then retired at age 58. "I just couldn't handle it anymore," he says. But he continues to oversee the cemetery decorating.

— Waldo Fisher
U.S. Army, Staff Sergeant, Company C, 3rd Battalion, 506th Infantry (Airborne), 101st Airborne Division
Vietnam: 1969–71

The Long Homecoming

More than 30 years after coming home from the Vietnam War, Douglas Haugstad helped bring another Soldier home as well. While on a humanitarian mission to the Democratic Republic of  Vietnam in 2000, Haugstad and his wife learned of a man who had found a wrecked, partially submerged helicopter. While sifting through the ruins for possible treasure, the man found teeth, bone fragments, and a dog tag of an American serviceman.

Now, years later, the man wanted to return the remains to the Soldier's family. Haugstad received a plastic sandwich bag that contained a charred dog tag with the name "Ritchey," and two brittle teeth.

"I was overwhelmed to think this could be a Soldier who had been missing in action all these years," Haugstad says, "but I didn't know if the dog tag was authentic or not, because in Vietnam, they had dog tags for sale in a lot of the shops."

His son, Lance Haugstad, began an Internet search and found the name of Lance Corporal Luther Ritchey Jr., a U.S. Marine from Ohio, still missing in action and presumed dead. In October 1963, Ritchey was aboard a UH-34D that crashed in mountainous terrain near Da Nang, and his remains—and those of his crewmember, Manuel Denton—were never found.

Had he lived, Ritchey would be the same age as Doug Haugstad, who served in Vietnam as a Signal Corps lineman, stringing electrical wiring for military units. Haugstad thought it would be a simple matter to get the remains identified. But in fact it took the help of a newspaper reporter, David Kranz of the (Sioux Falls, SD) *Argus Leader*, and U.S. Senator Tom Daschle from South Dakota, to finally get military officials to expedite the identification process. The teeth were sent to the Central Identification Laboratory in Honolulu, Hawaii, where they were eventually identified as Ritchey's.

In January 2004, Ritchey's 84-year-old mother, Dorothy Heslep, got the news that her son's remains had been positively identified. She wrote a letter of thanks to Doug and Lynne Haugstad that read, "After 40 years and eight months, I'm like a different person. The weight is lifted from my chest and I can breathe easier now. And I owe it to our God and his workers, like both of you, Mr. Kranz, the newspaper, and Senator Daschle. May God bless you all." Almost 41 years after he died, Ritchey's remains were buried with full military honors.

— Douglas Haugstad
U.S. Army, Specialist 4, lineman,154th Signal Battalion, Signal Corps
Vietnam: 1965–66

Our Suffering Is Mirrored on the Other Side

In 1994, more than two decades after he served in Vietnam, after he lost friends there and lost his older brother to cancer caused by Agent Orange, Lieutenant Colonel (ret) James G. Zumwalt visited Hanoi. He arrived full of anger, still convinced that the  war "was a black-and-white issue. These people had caused America great suffering. They had caused me and my family great suffering."

Zumwalt made the trip with his father, Admiral Elmo R. Zumwalt Jr., who had commanded the naval forces in Vietnam from 1968 to 1970. Their brother and son, Elmo R. Zumwalt III, had served in Vietnam in 1969 to 1970, and died in 1988. Elmo's death led the elder Zumwalt to investigate the medical repercussions of Agent Orange, used as a defoliant during the war to make it harder for the enemy to ambush U.S. troops.

The goal of the Hanoi trip was to enlist the Vietnamese government's help in a joint study about the effects of Agent Orange. What James Zumwalt hadn't counted on was that his heart would be changed in the process. On the third day of their visit, father and son met with Maj. Gen. Nguyen Huy Phan, who had served in the North Vietnamese medical corps during the war. Phan had also lost a brother during the war and was still haunted by the loss. The Zumwalts also visited a center for Vietnamese amputees.

"For the first time, I felt my heart go out to an enemy for whom I no longer felt hatred, realizing he too had suffered from war," Zumwalt wrote in his 2010 book *Bare Feet, Iron Will: Stories from the Other Side of Vietnam's Battlefields*. We have to realize, he says, "that our suffering is mirrored on the other side."

He eventually visited Vietnam 50 times, interviewing former North Vietnamese Army soldiers and Viet Cong. His goal in writing the book, he says, was not to glamorize the enemy but to humanize them.

"You are never going to help those having difficulty putting war behind them," he says, "unless you look at the enemy in human terms."

— James G. Zumwalt
U.S. Navy, naval officer, USS *Perkins* (DD-877)
Vietnam: 1969
U.S. Marine Corps, platoon commander, company commander, 1st Battalion, 4th Marines
Vietnam: 1971–72

A U.S. Navy Sailor of the River Patrol Force stands ready at his boat's dual .50-caliber machine guns. *Photo courtesy of DoD.*

The Monster Under the Bed

In 1994, 25 years after he served in Vietnam, Bill Ervin returned as a tourist. This time, he hoped, he would get to know the country and the people—and revisit the places that still haunted him.

He discovered that the hills along the DMZ where he had fought were still full of bomb craters deep enough to fit a house. The trees that Agent Orange defoliated were still bare. The musky smell of the jungle and the buzz of the insects triggered memories of what it was like to come face-to-face with mortality on the battlefield.

He began taking other veterans, and sometimes their wives and children, on trips to Vietnam. They visited the places where the men had trudged and fought and seen their buddies die. They also did touristy things: dinners and trips to the beach, sightseeing and shopping.

"Why would I want to go back to Vietnam—I see it every night in my dreams," some vets tell Ervin. He answers this way: "It's like when you're lying in your room as a kid and you're afraid there's a monster under the bed. And then you look under the bed and you see there isn't a monster, it's just your bedroom." The hard memories of the war will still be there, he tells them, but with a new perspective: "This is the past, and it's not controlling my life anymore."

In 2008, after his wife died, Ervin decided to teach English in Vietnam for a year. He's been there ever since, eventually marrying Nguyen Ahn, a former teacher and TV commentator whose family in Da Nang goes back 13 generations. Together they run a tour company, Bamboo Moon, and have built a house just minutes from China Beach, which is now called Non Nuoc. Things change; life goes on.

"Vietnam is much more than a war," says Ervin. "That's what I'm trying to get across to people." It's a strikingly beautiful country, filled with people who want to be friends. He lives in a neighborhood full of aging Vietnamese veterans. They tell him "the past is the past; we can look back at the sorrow of the past or we can look toward the future."

There are still reminders in the countryside that there was once a war there. But the underbrush has grown back and there is life. "The land is healing itself, even though scars remain under the green growth," Ervin says. "Kind of like myself."

— Bill Ervin
U.S. Marine Corps, Lance Corporal, Company D, 1st Battalion, 3rd Marines, 3rd Marine Division
Vietnam: 1969

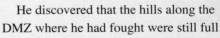

Soldiers of the 1st Cavalry Division (Airmobile) conduct Operation Pershing. *Photo courtesy of NARA.*

Traumatized refugees rescued from Saigon by a U.S. Marine helicopter move across the deck of aircraft carrier USS *Midway* (CVA-41) off the coast of South Vietnam on April 30, 1975. *Photo courtesy of DoD.*

Bookends to the War

Operation Frequent Wind—the plan to evacuate Americans and at-risk South Vietnamese from Saigon—began early in the morning on April  29, 1975. Ten time-zones away, Marine Major Robert "Bud" McFarlane was in the West Wing of the White House, where he was serving as military assistant to Secretary of State Henry Kissinger.

As the night wore on, McFarlane stayed on the phone with Graham Martin, U.S. ambassador to South Vietnam, and relayed live updates to Kissinger, who was shuttling back and forth to the Oval Office. By then, thousands of South Vietnamese were lined up around the embassy compound in Saigon, desperate to get on board the helicopters that would take them out to U.S. Navy vessels.

"The ambassador kept inflating the numbers [of refugees]," McFarlane recalls, "because he didn't want the operation to be turned off." More and more South Vietnamese ran to the embassy, hoping to escape. On the morning of April 30, 1975, Martin himself was reluctantly evacuated. Within hours, North Vietnamese tanks smashed through the gates of the South Vietnamese Presidential Palace and the war was over.

McFarlane, who later went on to become President Ronald Reagan's National Security Advisor, faults Congress for not providing continued support to South Vietnam after the Paris Peace Accords in January 1973.

He thinks of his own war efforts as "bookends" to the war. On one end was the March day in 1965 when he commanded a Marine artillery unit that landed at Red Beach 1 in Da Nang—the first landing of a U.S. combat unit in Vietnam. The other bookend was that April day, 10 years later, when Saigon fell.

McFarlane had been optimistic in 1965, certain that the U.S. and South Vietnamese military would be able to establish stability in a country reeling from decades of foreign occupation and internal division. As the last helicopter took off from the U.S. Embassy on the last day of April 1975, he felt what he calls "a searing sense of national loss."

— Robert "Bud" McFarlane
U.S. Marine Corps, Major, Assistant Secretary of State
Vietnam: 1965–66, 67–68
White House Staff: 1975

The North Vietnamese Army captures the city of Saigon on April 30, 1975, after South Vietnamese forces are overwhelmed and collapse. *Photo courtesy of NARA.*

The Final Hours

In the middle of the night on April 30, 1975, Foreign Service officer Ken Moorefield stood on the roof of the American Embassy in Saigon and counted the people in the courtyard down below. At least 450 South Vietnamese—including embassy employees and foreign diplomats—were gathered there, all desperate to flee their country as it fell to the North Vietnamese. He saw thousands more standing in the streets outside the embassy compound.

Who would escape and who would be left behind? Moorefield and his consular team had already spent two weeks struggling with this heart-wrenching dilemma. As an embassy consular officer, he had been working to orchestrate the screening and selection of thousands of American and Vietnamese civilians who sought to flee. Many had worked for U.S. government agencies, the South Vietnamese government, or various charitable agencies—and all risked imprisonment or death by the approaching enemy. It was the job of Moorefield and his team at the Tan Son Nhut Air Base evacuation center to decide who would be on the departing planes.

At first, he explains, they tried to follow the criteria set up by the U.S. State Department, "but there were just too many exceptions to the rule, and too little time." Many wanted to bring along their large extended families; some had fled from fallen cities farther north, arriving in Saigon without the necessary ID. It was left to Moorefield and his colleagues to decide their fate—and often that meant having to ask families to leave some members behind, to make room for others.

There was, for example, the Irish nun who showed up wanting to evacuate an orphanage full of Amerasian babies. South Vietnamese officials were embarrassed that orphans were being airlifted to America, and Moorefield worried that if the Vietnamese government got word that more babies were leaving, it would shut down the entire evacuation. So he came up with this plan: take the orphanage staff, he told the nun, and create paperwork to establish fake families for the orphans. Such "families" could be given the okay to leave.

The evacuation out of the airbase stopped abruptly on April 29, when the North Vietnamese Army attacked the base with rockets and artillery. With the last planes gone, Moorefield volunteered to guide U.S. military buses transporting people from predesignated pickup points to where they could be airlifted out by helicopter to waiting ships of the U.S. Navy 7th Fleet.

"Saigon was literally falling apart," Moorefield says. Cars crammed with families and baggage clogged the roads; throngs of people were trying to push their way onto the military buses, sometimes frantically stuffing babies through the windows. Moorefield eventually returned alone to the American Embassy, where helicopters were evacuating people from the parking lot and the roof. When it got too dark to land, he rigged up high-intensity lights. The helicopters continued to come in every 20 or 30 minutes, landing on the rooftop helipad.

At 4:30 a.m., an order came from the White House: the evacuation was over, except for U.S. embassy personnel. Moorefield explained to Ambassador Graham Martin, who had stayed to the end, that he'd been given a presidential order to board the next helicopter. Moorefield helped him get safely on board, then caught the next-to-last helicopter out, at about 5:30 a.m.

He can remember rising above the city: dawn was just breaking, and down below the city looked deceptively peaceful. But he knew what was coming: the North Vietnamese Army was on the outskirts of Saigon. "The whole country of South Vietnam was about to cease to exist," he recalls somberly. And he also knew that, despite their best efforts, time had run out for the many thousands of people who could not be saved.

— **Kenneth "Ken" P. Moorefield**
   U.S. Army, Captain, infantry, military advisor, Military Assistance Command, Vietnam (MACV),
   9th Infantry Division and 25th Infantry Division
   Vietnam: 1967–68, 69–70, 73–75

* After his service in Vietnam, Kenneth Moorefield remained with the State Department. In 2002, he was appointed ambassador to the Republic of Gabon and the Democratic Republic of Sao Tome and Principe. He is now Department of Defense Deputy Inspector General for Special Plans and Operations.

Sailors push overboard a South Vietnamese Air Force UH-1H used by fleeing South Vietnamese to escape the Communists. The small landing pads on the American ships are not large enough to accommodate the unexpected helicopters full of refugees. *Photo courtesy of DoD.*

Steaming Toward Safety

The helicopters came over the horizon of the South China Sea. One after another, scores of South Vietnamese UH-1 Hueys arrived, their doors and weapons removed to make room for families desperate to escape Saigon. Choppers built to hold eight passengers were crammed with 20, remembers Mike Thomas, a lieutenant on the destroyer escort USS *Cook* (DE-1083). On April 30, 1975, Thomas stood on the deck of the ship and watched the fleeing refugees.

Already, thousands of Vietnamese—at risk of imprisonment or death at the hands of the Communists—had been ferried out to U.S. Navy aircraft carriers. U.S. helicopters had evacuated them as part of Operation Frequent Wind. But these were South Vietnamese Army helicopters coming over the horizon—and nobody was expecting them.

"We had a small flight deck—we could only land one at a time—so as these planes started to show up, we land them, get the people out, and push the air frame over the side," says Thomas, who was the ship's anti-submarine warfare officer. This scenario was repeated on another destroyer escort, USS *Kirk* (DE-1087), as the day wore on, and on the aircraft carriers as well.

But that was only the beginning. That night, the destroyer escorts were sent to Con Son Island, where 30 Vietnam Navy ships, fishing boats, and cargo ships were waiting, jam-packed with refugees. On May 1, USS *Cook*, USS *Kirk*, and five other U.S. ships began to escort the motley Vietnamese fleet to safety. An estimated 33,000 Vietnamese were on those boats as they sailed toward the Philippine Islands.

To ensure coordination, Thomas was assigned to a Vietnam Navy ship, *Ngo-Quyen*. A ship meant for a crew of 150, it was now bursting at the seams with about 1,000 people. "Every square inch of deck space had somebody sleeping on it," Thomas recalls. "Cleanliness, food, and water were key concerns." The food arrived via U.S. Navy ships—C-rations and pallets of bagged rice—and American Sailors would come aboard periodically to wash the decks down with fire hoses to help prevent the spread of disease.

But as the ships steamed east, a new obstacle arose: Philippine president Ferdinand Marcos, reluctant to alienate the victorious North Vietnamese government, refused to let South Vietnamese vessels enter port. The solution was the official transfer of ownership of all refugee ships over to the U.S. Navy—which meant taking down each South Vietnamese flag and raising an American flag in its place.

There was a hastily devised flag ceremony. "It was very somber," Thomas remembers. "And simple, in that nobody had access to a lot of 'pomp and circumstance.'" At least one U.S. Navy member was assigned to raise the U.S. flag on each ship. Thomas performed this duty on *Ngo-Quyen*.

Before transferring their ship to U.S. custody, officers and men of the South Vietnamese Navy stand at attention as their nation's flag is lowered for the last time. *Photo courtesy of Liem Bui.*

For the South Vietnamese Navy and Army officers on board, it was a sad, symbolic end to the country for which they had so bravely fought. "But they knew they were fortunate," says Thomas, "to have gotten out."

— **Michael C. Thomas**
 U.S. Navy, Lieutenant, anti-submarine officer, USS *Cook* (DE-1083), 7th Fleet
 Vietnam: 1971, 1973, 1975

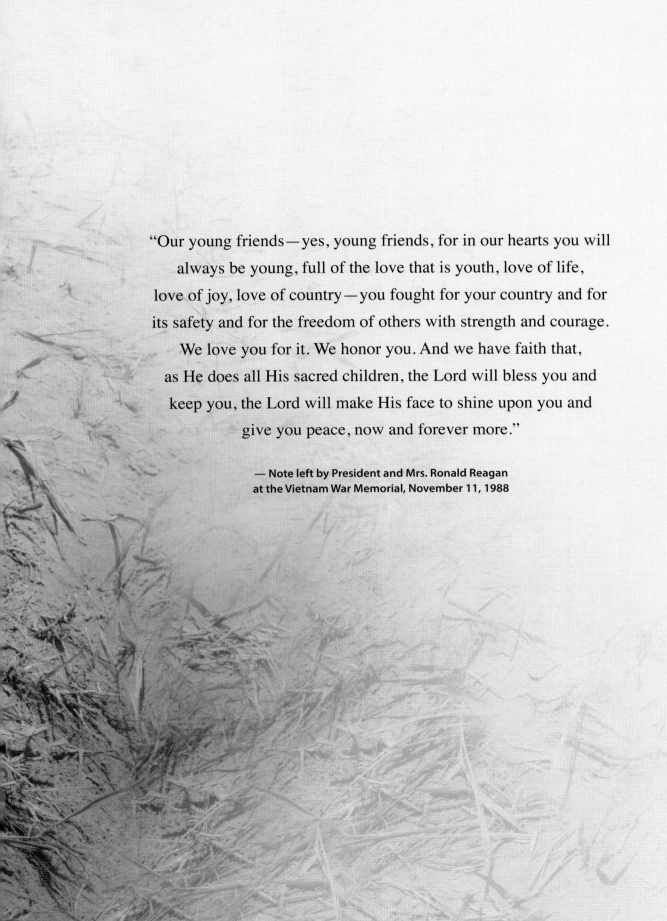

"Our young friends—yes, young friends, for in our hearts you will
always be young, full of the love that is youth, love of life,
love of joy, love of country—you fought for your country and for
its safety and for the freedom of others with strength and courage.
We love you for it. We honor you. And we have faith that,
as He does all His sacred children, the Lord will bless you and
keep you, the Lord will make His face to shine upon you and
give you peace, now and forever more."

— Note left by President and Mrs. Ronald Reagan
at the Vietnam War Memorial, November 11, 1988

PUBLISHING CREDITS

Publisher: **John Lund**

Managing Editor: **Sharlene Hawkes**

Advising Editor: **Colonel (ret) Raymond K. Bluhm Jr., U.S. Army**

Production Director: **Daryl Guiver**

Art Director: **Darren Nelson**

Senior Editor: **Elayne Wells Harmer**

Senior Writer: **Elaine Jarvik**

Writers: **Kellene Ricks Adams**
Valerie Phillips
Rebecca Thomas
Caroline Lambert
Nancy Van Valkenburg

Consultants: **Colonel (ret) Richard Kiernan, U.S. Army**
Dr. Edward J. Marolda

Research Assistants: **Ike Hall**
Brandon Young
Renee Casati
Peggy Mitchell
Dan Evans
Nancy Allred

Photo Researcher: **Gina McNeely**

Special Photo Contributor: **Russell A. Elder**

THANK YOU

This publication was made possible thanks to the generosity and support of our founding partners:
Lieutenant Colonel (ret) E. W. "Al" and Kathleen Gardner | Joe and Kathleen Sorenson

Thank you to the following individuals for their invaluable support :
Lee Allen | Leslie Beavers | Rosie Berger | John Beerling | Curt Bramble | Todd Creekman
Elizabeth Dole | Jim Fewell | Jim Fisher | John Garcia | Heather French Henry | Jim Holbrook
COL James P. Isenhower III | Rick Kiernan | Jim Knotts | Fred Lyons | Robert McFarlane
Monica Mohindra | John Muckelbauer | Bob Patrick | Randy Reeves | Julie Rodriguez
John Rose | Jon Voight | Bob Wallace | Lonnie Wangen | Dave Winkler

Thank you to the following partners:
Veterans of Foreign Wars | Vietnam Veterans Memorial Fund | Naval Historical Foundation

Thank you to the Remember My Service Productions Advisory Board:

Honorable William Chatfield, U.S. Marine Corps
Major General (ret) Pat Condon, U.S. Air Force
Major General (ret) Peter S. Cooke, U.S. Army
Sergeant (ret) Sammy Davis, Medal of Honor Recipient
Colonel (ret) Jim Fisher, U.S. Army
Command Sergeant Major (ret) John Gipe, U.S. Army
Captain (ret) Dale Lumme, U.S. Navy
Brigadier General (ret) Anne Macdonald, U.S. Army
Major General (ret) John Macdonald, U.S. Army
Chief Executive Officer Angela Phillips, Middletown Tube Works
Daniel "Rudy" Ruettiger, Vietnam Veteran
General (ret) Walter L. "Skip" Sharp, U.S. Army

Special appreciation to:
National Association of State Directors of Veterans Affairs (NASDVA)
National Conference of State Legislatures (NCSL)
U.S. Library of Congress Veterans History Project
The Vietnam Center and Archive, Texas Tech University
Veterans of Foreign Wars
The White House

and most especially

The many Veterans who contributed
their stories to this project

For more stories, visit the archives:
www.Vietnam50gift.com

Stories Directory